KEEPING A
GEC
By Rol

Northern Giant Cave Gecko *(Pseudothecadactylus)* **photo by Gunther Schmida**.

I would like to thank Darren Green and Jason Goulding for initiating this publication and coordinating its production. Also thanks to Darren Green, Janet Cheriton, John Rudge and Joanne Carnsew for taking the time to read the first draft and making constructive comments and suggestions. Several photographers have made great contributions to this book and I am very appreciative of their exceptional work which has played a vital role in producing the finished product, each has been credited with their photographs throughout the book. I would also like to express my gratitude to a plethora of gecko keepers in Australia, Europe, New Zealand and the USA who have all helped to promote the keeping of these reptiles and assisted me in many ways over the last 30 years. Finally a big thanks to my wife and daughter for putting up with me and my hobby over the years and encouraging my enthusiasm in these wonderful creatures.

Australian Reptile Keeper
PUBLICATIONS

TITLE: Keeping Australian Geckos.
AUTHOR: Rob Porter.
PHOTOGRAPHS: Sarah Codman, Jason Goulding, Darren Green, Barry Kiepe, Kelvin Marshall, Byron Manning, Rob Porter, Gunther Schmida & Troy Webb.
ILLUSTRATIONS: Jason Goulding & Rob Porter.
PUBLISHER: Australian Reptile Keeper Publications.
EDITORS: Jason Goulding & Darren Green.
EDITORIAL CONSULTANTS: Joanne Carnsew. Janet Cheriton & John Rudge.
DESIGN & LAYOUT: Jason Goulding.
PRINTING: The Printing House, Adelaid SA email tphprint@ozemail.com.au
ISBN: 978-0-9758200-2-5

© **COPYRIGHT 2008** Reproduction of this publication in whole or part, including photographs, is not permitted without prior written consent of the publisher. All rights reserved.

Publisher's note...

Australian Reptile Keeper Publications© is providing a series of basic care books for reptile and amphibian enthusiasts, the group of people we call herpetoculturalists. The series is aimed to encourage reptile and amphibian keeping by giving the novice access to the experience and advice of leading herpetoculturalists. These books, written by enthusiasts with considerable knowledge and practice in their field, are based on the author's experience and are not necessarily the sole, or absolute, method of keeping. We simply hope to assist the reader with a general understanding of the herp and provide sufficient information on basic husbandry. A reference section is provided for the reader who requires additional information.

In Australia, taking reptiles and amphibians from the wild is an offence. Wildlife regulations are also enforced, and the keeping of these reptiles and amphibians may require a licence or permit. Before obtaining any herps, check with your state's wildlife regulations and ensure you are purchasing an animal from a reliable source. Some of the reptiles covered in these books may be dangerous and contact with such animals, without the proper experience, regard to safety, or medical attention could result in serious injury or death.

Australian Reptile Keeper Publications© welcomes original manuscripts, including photographs and illustrations, on all herp related topics. Submissions should be directed to Australian Reptile Keeper Publications©, PO Box 2189, Bendigo MC Vic 3554,
e-mail info@reptilepublications.com.au or www.reptilepublications.com.au

Australian Reptile Keeper Publications© encourages all reptile and amphibian enthusiasts to belong to a club or society. Not only do you get to meet other people with a similar interest, but you can also benefit from the pool of knowledge that goes hand in hand with belonging to a club. For a list of clubs and societies closest to you, write to Australian Reptile Keeper Publications© and include your field of interest. Remember, informed keeping is a responsibility that may be enjoyed for many years.

CONTENTS

1. **INTRODUCTION**
1. What are geckos?
2. About this guide
2. Morphology & ecology
8. Australian geckos & legless lizards
13. Why keep geckos?
14. Acquiring a gecko
15. Quarantine
15. Do I need a licence?
16. Choosing a gecko
17. Handling geckos

19. **HOUSING**
19. Enclosure construction
24. Heat & light
26. Moisture & humidity
28. Substrate
30. Enclosure furnishings

33. **NUTRITION**
33. Food
36. Supplementation

37. **BREEDING**
37. Sexing geckos
40. Mating & egg laying
44. Egg Care
46. Hatchling care

50. **HYGIENE & HEALTH**

51. Calcium/D3 Deficiency
52. Mites
52. Gut impaction
53. Sloughing Problems
54. Egg binding

SPECIES LIST.
55. Helmeted gecko
(Diplodactylus galeatus)
57. Golden-tailed gecko
(Strophurus taenicauda)
59. Spiny-tailed gecko
(Strophurus ciliaris)
61. Thick-tailed gecko
(Underwoodisaurus milii)
63. Common knob-tailed gecko
(Nephrurus levis)
65. Centralian knob-tailed gecko
(Nephrurus amyae)
67. Giant cave gecko
(Pseudothecadactylus cavaticus & lindneri)
69. Marbeled velvet gecko
(Oedura marmorata)
71. Southern spotted velvet gecko
(Oedura tryoni)
73. Sydney broad-tailed gecko
(Phyllurus platurus)
75. Rough-throated leaf-tailed gecko
(Saltuarius salebrosus)
77. Bynoe's gecko
(Heteronotia binoei)
79. Ring-tailed gecko
(Cyrtodactylus louisiadensis)
81. Common scaly-foot
(Pygopus lepidopodus)
83. Burton's legless lizard
(Lialis burtonis)
85. References

INTRODUCTION

What are Geckos?

Technically, geckos are small lizards belonging to the family Gekkonidae with a wide distribution that is most diverse in the tropics. To many, however, geckos are much more than a bunch of scaly creatures. They invariably invoke an endearing quality unequalled amongst other reptiles. Their appeal lies in their agreeable appearance highlighted by huge jewel-like eyes and velvet skin. Their well-developed feet and toes are apparently out of proportion to the rest of the body yet these make many geckos consummate performers in the arts of climbing and agility. In addition, unlike many of their reptilian kin, many have accepted humans and their buildings as part of their own world and competently perform an invaluable function in pest control. This partnership has developed to the extent that it is considered a good omen in many areas of Southeast Asia if a tokay gecko *(Gekko gecko)* decides to take up residence in your home. Some species of gecko have become so attached to human habitation that they are rarely found elsewhere and have proceeded to follow mankind through his travels around the world by hitchhiking amongst his cargo and freight and rapidly establishing in new habitats on arrival.

Geckos also have an unlikely close relative. The legless lizards or pygopods resemble snakes more than geckos but are in fact classified in the same family. Numerous morphological (the form of the body), ecological (their life history) and behavioural similarities connect the two groups in the reptile family tree and for this reason both have been included in this publication.

Despite the relationship that has developed between these reptiles and humans, the gecko's secretive and mostly nocturnal habits have until recently meant these fascinating and appealing lizards have been poorly understood. There is still much we do not know about how these seemingly delicate reptiles live and survive in some of the harshest climates in the world. Their climbing feats

Barking or thick-tailed gecko *(Underwoodisaurus milii)*.

Photo by Byron Manning.

are still a topic of much debate as to exactly what the mechanical nature of this ability really is. The answers to this question may well provide valuable assistance for human technological developments. As we look more closely at this large group of lizards there will undoubtedly be much more that we will learn further enhancing our appreciation of these animals.

About This Guide

This guide is designed to provide sufficient basic information to assist those with little or no experience of keeping geckos to successfully maintain and breed many species of Australian geckos. Information is provided on all aspects of basic husbandry techniques. However, there are some aspects of husbandry that are outside of the scope of this publication such as the identification and treatment of disease and specialist husbandry techniques for more difficult species. For those interested in these and other issues not dealt with in depth in this book, there is a plethora of literature published overseas and on the internet, although all information should be verified from a variety of sources before being implemented and specific health aspects should be addressed in consultation with a veterinary surgeon. While the legless lizards are also included in this publication, the extent of our knowledge concerning this group is very limited. Consequently, the majority of the information will relate to the true geckos with legless lizard data provided where it is available.

There is still much we do not know about many areas of reptile husbandry and geckos are certainly no exception to this rule. I cannot stress how important it is for keepers to observe record and then disseminate this information by publishing the details. It makes no difference whether this data is reproduced in a local society newsletter or a professionally printed magazine but it is important to ensure that your unique records of reptile behaviour, breeding, feeding and other observations are accessible to others. This way the hobby will continue to progress, our overall knowledge and understanding will grow and our ability to provide a high standard of care will keep improving.

Morphology and Ecology

Geckos as a group have been around for some 50 million years. Currently, there are almost 1200 recognised species worldwide making the family second only to the skinks in size. More recently the group of legless lizards *(Pygopodidae)* have also been included in the gecko family. According to the most recently published guide (Wilson & Swan 2008), Australia has 115 species of gecko and 41 species of pygopods. True geckos are small to medium sized lizards ranging in size from the giant New Caledonian gecko *Rhacodactylus leachianus* that reaches over 350mm in total length, down to a diminutive Caribbean Island species of *Sphaerodactylus* that rarely exceeds 35mm. In the recent past geckos grew to a much larger size than existing species. One New Zealand gecko *Hoplodactylus delcourti* is represented as a single preserved museum specimen measuring around 600mm in total length. Australian geckos range from the impressive leaf-tailed species *(Saltuarius)* exceeding 250mm in total length to the clawless gecko *(Crenadactylus ocellatus)* of central and Western Australia with a length of around 60mm. By contrast the legless lizards can attain 1 meter in overall length, the largest being the common scaly-foot *(Pygopus lepidopodus)*, the smallest is the collared delma *(Delma torquata)*, which is about the diameter of a shoelace and 200mm in length.

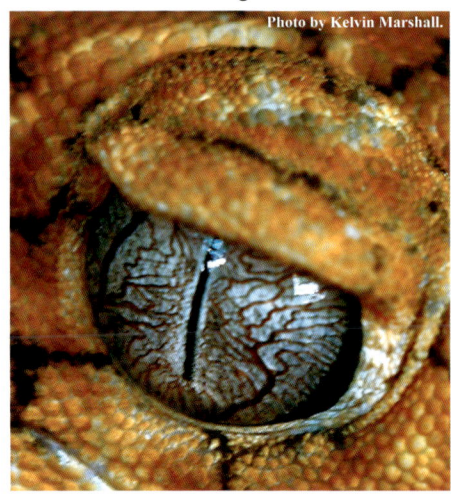

Close-up of gecko eyes, northern leaf tail gecko *(Saltuarius cornutus)* **left, Centralian knob-tail gecko** *(Nephrurus amyae)* **right.**

Although geckos are renowned around the world for their superb climbing ability, this is not the case with all species. Many geckos, particularly desert dwellers, have no need for such aptitude because of their terrestrial lifestyle and their toe structure has evolved accordingly. Indeed the toe morphology of geckos is so varied that it has become an important tool for identifying closely related groups within the gecko family. The granular scales that cover the body produce the velvety texture of the gecko skin. In some species these are all fairly uniform in size, while others may possess enlarged or raised scales on various parts of the body. Tail form and function is also highly variable. Nowhere is this more extreme than in the knob-tail *(Nephrurus)* and leaf-tail gecko *(Saltuarius* and *Phyllurus)* species. In some species of knob-tailed gecko the tail is reduced to a tiny spherical ball, while the latter group have tails that are highly elaborate and ornamented. Most species still have the ability to drop their tails if threatened, but with species such as the leaf-tailed geckos there is only a single fracture point so the whole of the tail is dropped rather than having the option of shedding a small portion, which is available to most other species.

One of the most striking features of many geckos are the eyes. These are always large and well developed; an indication of both the gecko's nocturnal habits and their reliance on vision to locate and capture food. Many species have highly patterned and coloured eyes that resemble precious jewels in finely textured settings. The vertical pupil is dark in colour and is reduced to a narrow slit in bright light, yet broad enough to cover most of the eye in the dark thus collecting as much light as possible providing geckos with unequalled nocturnal vision. The majority of species, including all Australian geckos and all pygopods, have no eyelid, meaning the eye cannot be closed and is therefore also difficult to keep clean. These lizards have solved this problem by developing a long, flat, fleshy tongue that is used to lick the eyeballs clean.

Geckos and legless lizards are the only lizards that produce sound and can communicate by vocalization. Most calls are simple squeaks and clicks, while others are a little more elaborate such as the Tokay gecko *(Gekko gecko)* whose name was derived from its familiar call. The ability to vocalise is one of the characteristics that indicate a close relationship between the true geckos and the legless lizards despite their apparent dissimilarity in physical appearance. Other features in common include the use of the tongue to clean the eye, similar reproductive strategies and numerous internal morphological features.

Geckos have proved to be extremely adaptable reptiles and have colonized a huge range of habitats from desert and savannah to rainforest and coastal heath. Indeed, the ability of some geckos to occupy a range of habitats has seen some species become the most widely distributed of all the lizards. Rafting by natural means or hitchhiking on human transport are techniques used by several species to extend their range. Several species have made landfall in Australia in this fashion and one, the Asian house gecko *(Hemidactylus frenatus)* is rapidly spreading through the tropical and subtropical parts of the country. What effect this may have on native species remains to be seen.

Unlike a small number of overseas species, all Australian geckos are primarily nocturnal in habit. The peak time of activity usually occurs within the first two to four hours immediately after sunset. The lizards emerge from their daytime retreats and utilize their large and highly sensitive eyes to locate living prey, mostly in the form of invertebrates. Some species, such as the dtellas *(Gehyra* spp.*)* and velvet geckos *(Oedura* spp.*)* are active hunters that stalk and pursue their food. Others, such as leaf-tailed geckos, are ambush feeders and employ a sit and wait tactic, remaining immobile for hours until an unwary insect crosses their path. The extent of nightly activity periods is highly dependent on prevailing temperatures. On cooler nights, geckos may only emerge for a very short period of time and do not leave the immediate vicinity of their day retreat, while in winter activity may cease altogether for many weeks. Other environmental conditions may also affect nocturnal activity. Excessive wind, rain or even a full moon in a cloudless sky may restrict or even inhibit active periods in some species.

The legless lizards vary greatly in their activity patterns. Many species are diurnal while others are predominately nocturnal in habit. Other species seem to exhibit classic crepuscular periods of activity, that is during the early morning and late afternoon/evening, thus avoiding the extreme temperatures of both night and day. In some species, activity periods may change seasonally with normal daytime activity being replaced by nocturnal habits during the hottest part of the year.

The majority of Australian geckos and pygopods are generalized insectivores in their feeding habits. This means they will eat almost any insect or other small invertebrate that they are able to catch and overpower. The jaws of the larger species are very strong enabling them to crush the outer shells of large beetles and cockroaches with ease. The common scaly-foot *(Pygopus lepidopodus)* is somewhat of a spider specialist and there are reports of them catching and eating even the deadly funnel-web spiders. Some small geckos, especially desert inhabitants, have become quite specialized on termites as food items and can be difficult to persuade to accept any other types of insect prey. A few species, such as the Centralian knob-tailed gecko *(Nephrurus amyae)* will also go for vertebrate prey in the form of their smaller gecko cousins. Although numerous overseas species of gecko have been recorded feeding on the nectar of flowering plants, this activity appears to be rare in Australian species. although there have been some records of species feeding on tree sap. In fact it is the legless lizards that seem to have developed more of a sweet tooth. Many species will eagerly accept small ripe berries and fruits and the brigalow scaly-foot *(Paradelma orientalis)* has developed a particular taste for *Acacia* sap.

Most geckos are intolerant of the high temperatures that are the preferred choice of dragons and goannas. Indeed, consistent temperatures in the mid-thirties would cause the death of some cold adapted species such as the barking gecko *(Underwoodisaurus milii)* and rainforest inhabiting leaf-tailed species. However, geckos still adjust their behaviour to increase or decrease their body temperature, a process called thermoregulation. Raising the body temperature is usually carried out in an indirect manner by using other objects warmed by the sun, such as the underside of an exposed rock, rather than direct exposure. Some species may bask, though this is probably achieved in a cryptic manner with only a small part of the body exposed to the sun's rays while the majority of the lizard is still hidden in a retreat. At the other extreme, some species are highly tolerant of cold temperatures and will remain active and feeding at temperatures well below 15°C.

Reproduction in geckos and pygopods is remarkably uniform producing two, or rarely one, egg or baby in each clutch. Most species are egg layers, only the New Zealand geckos and one species in New Caledonia producing live young. The majority of native species produce eggs with a parchment or flexible shell, while others such as the ring-tailed gecko *(Cyrtodactylus louisiadensis)*, the dtellas *(Gehyra* spp.*)* and Bynoe's gecko *(Heteronotia bynoei)* lay hard-shelled or calcareous eggs more similar to a bird's egg. Field studies indicate most species take between two to four years to reach sexual maturity and breed for the first time. In most species mating occurs in spring with eggs laid 3 to 5 weeks after. Multiple clutches are common in captivity in many species and may also arise in wild geckos when conditions are favourable. There is usually about a month or so between each clutch, although a fresh mating is not always necessary as some female geckos can store sperm. Some species are communal egg layers with multiple females depositing their clutches together in a favoured spot, often over many seasons. The populations of a small number of Australian geckos in the genus *Heteronotia* are all females, and these animals are able to reproduce without males, a process known as parthenogenesis. They are currently all described as Bynoe's geckos but it is thought that this is a complex of several distinct species, hybridisation of some having produced the parthenogenic groups.

Photo by Barry Kiepe.

Centralian knob-tailed gecko *(Nephrurus amyae)* **displaying threat behaviour.**

Some Australian leaf-tailed geckos such as this northern leaf-tailed gecko *(Saltuarius cornutus)* are exquisitely camouflaged and blend perfectly into their background complete with patches of lichen and bark texture

Australian Geckos & Legless Lizards

Australia is endowed with some of the most unusual and fascinating species of gecko as well as the majority of species of legless lizard. Of the over one hundred species of true gecko in this country; some are unique and immediately recognisable such as the endearing knob-tailed geckos *(Nephrurus)*. There are nine species in this group ranging from the dainty smooth knob-tail *(N.laevissimus)* to the robust rough-skinned Centralian knob-tail *(N.amyae)*. Many make good captives and have become very popular because of their strange appearance characterised by short tail terminating in a round knob. At the other extreme the leaf-tailed geckos *(Saltuarius and Phyllurus)* possess a truly impressive tail. It is broad and highly ornamented with spines and tubercles mimicking the shape of a leaf and helping to break up the lizard-shape outline to passing predators. Numerous new species of leaf-tailed gecko have been described in recent years. It appears that some populations have become isolated on several north Queensland rainforest-covered mountain tops and have formed new species in the absence of genetic exchange with other populations. It is likely that there are still more species in similar situation awaiting discovery and naming.

Another unique group of Australian geckos is the tail squirting members of the genus *Strophurus*. These inhabitants of the dry inland areas of the country react to threats by squirting a black sticky substance from glands within the tail that deters likely predators. Other species of gecko have different means to warn off attackers. Some like the barking gecko *(Underwoodisaurus milii)* carry out distinctive behavioural acts such as raising themselves onto their toes, arching the back and waving the tail like a flag. This can often be combined with an open-mouth lunge towards the threat, combined with a rasping squeak. All of these performances can often be seen in captive specimens.

A stone gecko *(Diplodactylus vittatus)* exhibiting threat behaviour by opening the mouth and waving the raised tail from side to side.

Many Australian geckos are beautifully marked and coloured, especially as juveniles. Young reptiles are often more brightly coloured than the adults and nowhere is this more evident than in the velvet geckos *(Oedura)*. While the adults of species such as the northern velvet gecko *(O.castlenaui)*, the marbled velvet gecko *(O.marmorata)* and the fringe-toed velvet gecko *(O.filicipoda)* are attractive reptiles in their own right, the hatchlings of these species are almost fluorescent in their colours. Often the colour and pattern of these youngsters bears little or no resemblance to the appearance of the mature animal. The situation is reversed in the golden-tailed gecko *(Strophurus taenicauda)*, which is predominantly a dirty grey colour when first hatched. Over time, the intricate reticulated pattern of black over pale grey develops and the stunning bright orange tail stripe and eye become more and more prominent.

Many Australian geckos have the ability to change the shade of their overall colouration to either match their environment and/or their behavioural mood. This is not the same as the rapid colour changes seen in lizards such as the chameleons, however, the effect is still quite dramatic in some species. When at rest most geckos are relatively dark and their markings do not stand out boldly.

Adult and hatchling colouration of the northern velvet gecko *(Oedura castlenaui)*.

Australian Reptile Keeper

This is probably an adaptation to blend into their resting environment and thus reduce their vulnerability to predators. When active at night, the background colouration of some species often becomes paler contrasting with the other markings and making the pattern appear much stronger e.g. the golden-tailed gecko. Other species, such as some of the dtellas *(Gehyra* spp.*)* may become paler all over at night with some species appearing almost translucent in appearance. The camouflage colouration of some species is truly amazing, particularly in some of the large leaf-tailed geckos *(Saltuarius* spp.*)* where they take on the intricate patterns of tree bark complete with patches of moss and lichen. Their vanishing trick is completed with the unusual broad tail breaking up the body outline and the entire upper surface and sides being covered with pointed spines and tubercules giving the lizard the rough textured appearance of its surroundings.

The dtellas *(Gehyra* spp.*)* are a familiar group of species, especially in country Australia. Some have become popular inhabitants of human dwellings and carry out an important task in controlling pest insects. The dtellas have been replaced by the introduced Asian house gecko *(Hemidactylus frenatus)* in many of Australia's tropical and subtropical cities. This little grey-brown gecko has reached our shores by hitchhiking on freight transported by sea and air and has quickly spread into the suburbs. Whether this invasion has been at the expense of the native dtellas is unclear. It is possible that the native geckos had already disappeared from built up areas, being unable to cope with the change of habitat and degree of disturbance and the introduced alien has then simply filled the vacant niche.

The legless lizards, while more uniform in their morphology, are still interesting and often attractive reptiles. Burton's legless lizard *(Lialis burtonis)* occurs in a huge range of colours from brick red to brown, grey and cream often with stunning white and dark brown racing stripes along the side of the head. Similarly, some of the scaly-foot group *(Pygopus)* may possess intricate patterns and markings, sometimes resembling the colouration and markings of some of the venomous brown snake species *(P.seudonaja)*. This appearance is often further enhanced by threat postures that also mimic these dangerous snakes. One amazing fact about this group is that about three-quarters of the length is made up of the tail. Consequently, if the tail is lost during an attack the stumpy little body that is left is a very sorry sight until the tail regenerates. There are some 41 species of pygopod in Australia though by far the majority of these have rarely if ever been maintained successfully in captivity for any length of time. Clearly, there is a great deal about this group that we have still to learn.

Many Australian geckos are blessed with unusual and sometimes weird tails.

Spiny-tailed gecko *(Strophurus ciliaris)*, **complete with glands able to squirt a sticky substance in defence.**

Thick-tailed gecko *(Underwoodisaurus milii).*

The strange knob of the knob-tailed gecko *(Nephrurus* sp.*)* is of unknown function.

The broad, fleshy tail of the fringe-toed velvet gecko *(Oedura filicipoda*) acts as a fat storage device for times when food is scarce.

The beautifully ornamented tail of the leaf-tailed gecko *(Saltuarius* sp.*)* is designed to assist the gecko with blending into its environment to avoid predators.

Australian Reptile Keeper

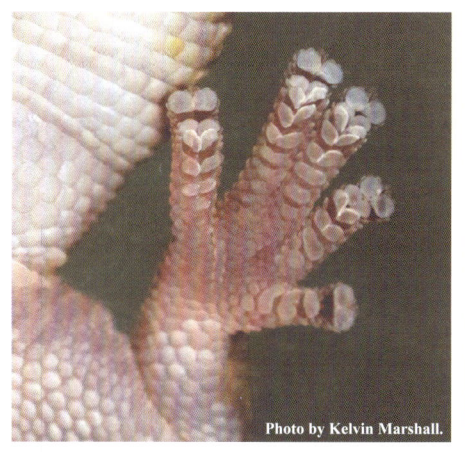

Examples of the variety in the form of Australian gecko feet. Clockwise from top left the toes of the velvet gecko (*Oedura* sp). have pads along most of its length for general climbing. The giant cave gecko *(Pseudothecadactylus cavaticus)* has broadened digits with very efficient pads for advanced climbing abilities. Centralian knob-tailed geckos *(Nephrurus amyae)* do not climb therefore only have small claws at the end of each toe. The toes of golden-tailed gecko *(Strophurus taenicauda)* and its relatives have pads only at the end reflecting the varied lifestyles of the group which spend time on and above the ground.

Why Keep Geckos?

In today's busy world, time and space have become important commodities. The increase in popularity of reptiles as pets is due in no small amount to this situation. Many are low maintenance animals, especially when compared with more traditional pets like mammals, birds and fish. Geckos are ideal in this respect. Their size requires small and basic housing, they are undemanding in heat, light and food, they are hardy, easy to breed and long-lived and they are not noisy, smelly or aggressive yet they are still attractive, active and fascinating creatures. What's more, they are ideal for reptile keepers who work full time, as they are active and visible at night when many other reptiles are sleeping. Truly, the perfect pet!

From a scientific point of view, the appropriate captive maintenance of these animals provides an insight into their secretive lives. While field research is an essential element in understanding the natural history of a species, accurate records of captive animals can also contribute important facts and data particularly with small, cryptic species that have little economic importance such as geckos and legless lizards. A well-maintained record keeping system is a must for any keen hobbyist to ensure the correct information is recorded and ultimately disseminated to others. Important data such as growth rates, age to maturity, breeding frequency, incubation times, etc. are all relatively easy to obtain from captive reptiles but can be difficult and often time consuming to extract from field research. Once this information has been acquired, publication is the essential next step to allow others access to the information. This doesn't have to be of scientific standard but its facts must be accurate. It also doesn't

Golden-tailed gecko *(Strophurus taenicauda)*.

matter if the information has been published elsewhere before, as a new angle on the same subject can often shed light on areas not previously examined, or provide a different perspective on unsolved problems or little known aspects of life history.

Acquiring a Gecko

The popularity of geckos as captive subjects is a relatively recent trend in Australia. Despite widespread interest in the group overseas, local hobbyists have ignored these lizards in the past in favour of the larger but more time-consuming goannas and dragons. For this reason the number of geckos captive bred in Australia is still low and this may well make it difficult tracking down available specimens. Geckos only produce one or two eggs per clutch and two to four clutches a year so captive numbers increase fairly slowly. However, a few specialist breeders produce numbers every year and their hardy nature and ease of breeding means more geckos are available each year.

The first step in obtaining a gecko should be to decide which species is the first preference. This choice goes hand in hand with research. Find out as much as you can about the species that interest you, not only their captive needs but their natural history as well. This information will prove invaluable when it comes to providing the proper care and housing for your lizard. Many publications are available with information concerning geckos particularly overseas. While these may not always deal with Australian species, much of the information is applicable to native geckos. There are also plenty of websites dealing specifically with geckos, however, as with any topic do not believe everything you read on the web and make sure you verify details you are not sure about from other sources. There is an excellent international group of gecko specialists and enthusiasts called the Global Gecko Association (www.gekkota.com), which produces a first class journal and a newsletter full of invaluable information. Many of the journal articles cover Australian species, often written by overseas keepers with many years of experience with these reptiles.

While most species are low maintenance captives some are more straightforward than others. Velvet geckos *(Oedura)* for example are an ideal beginner's gecko. They are very hardy and undemanding in their caging, captive environment and food requirements and many will grow rapidly and breed in their second or even first year. On the other hand some of the knob-tails *(Nephrurus)* can be a little fussy about their microhabitat and their food and may need a little more care and attention. Once the choice has been made, the next step is to organise the housing. Always ensure you have your enclosure ready to go before you obtain your gecko. A newly acquired lizard is always stressed from being handled and removed from familiar surroundings, so this condition must not be compounded by using substandard temporary accommodation.

Quarantine

Once the lizard has been purchased and brought home it should ideally be placed in its intended long-term housing. This reduces the stress on the animal, which can be significant if it is necessary to catch and move it again shortly after. In some situations this policy is not possible and temporary housing is required. This will be needed if the gecko is to be housed with other already established specimens. In this case it is important that the new arrival is maintained in a quarantine situation for at least a few weeks to ensure its good health before it is introduced to the others. This is particularly important if internal or external parasites have been recorded. These should be treated and eradicated where possible before contact with other lizards is permitted. This will also allow the new lizard to become accustomed to the new husbandry techniques before the potentially stressful experience of being introduced to a new enclosure with its established inhabitants.

A quarantine enclosure should be very simple and easy to clean. Use paper towel on the floor rather than any natural substrate and keep furnishings to a minimum. This will permit clear observation of the new arrival and quick and easy hygiene levels to be maintained at all times. Monitor closely the reptile's activity and feeding patterns and check that the droppings are well-formed dark pellets not watery and messy. If the latter occurs internal parasites may be causing the problem and samples should be taken to a vet for analysis.

Do I Need A Licence?

All Australian state authorities now have licensing systems in place requiring everyone to obtain a licence before acquiring most reptiles including geckos. Contact your local Environmental Protection Agency or National Parks office to obtain the information you require about the system in operation in your home state. Most gecko species are listed on the lowest category of permit in all states. However, some species have been allocated higher ratings because of their conservation status or the level of husbandry skill necessary to look after them. Once a licence has been issued to keep geckos, the lizards must then be obtained from another licensed keeper either from within your state or interstate. An import/export permit is usually required for the latter transaction.

There are several options open to locate a breeder of geckos. Firstly join your local herpetological society. This will put you in touch with like-minded people who share your interest and may often be good contacts for others who can supply animals, as well as an important source of information and experience on the maintenance of all reptiles. Secondly, check out the Australian websites that advertise reptiles for sale or exchange. Remember these have to be Australian sites as it is illegal for private individuals to import reptiles into Australia. Finally, some states permit public advertising of reptiles for sale, so keep an eye on the pet sections of your local newspaper.

Choosing a Gecko

When choosing any pet it is important to observe the animal when it is active to assess if it appears bright, alert and healthy. The nocturnal habits of geckos may make this potentially difficult. Needless to say, it is essential that you can observe the lizard carrying out its normal behaviours before making a decision to purchase and if this means carefully disturbing the animal from its daytime retreat then so be it. As long as the gecko displays bright eyes that are clear and not weepy, it moves without any obvious hindrance, its skin does not seem wrinkled or have pieces of old skin still attached to its head or toes and its tail condition looks good, i.e. the tail is full and rounded at its base not spindly and sunken the animal should be a safe purchase. A tail that has been lost and has regenerated is obviously different from the original tail in most species. However, a regrown tail itself is not a sign of an animal in poor health. As long as the new tail is full and without kinks and the lizard's overall health appears sound, a regenerated tail does not compromise condition.

Be observant. Look carefully at the lizard's housing. Is it clean and appropriate in size, shape, etc. to house that particular reptile? Ask questions about the animal, its behaviour, history and husbandry. This will not only give you an insight into how it has been maintained but may also assist you with its future care. Most geckos do not enjoy being picked up, so keep handling to a minimum. Simply watching the movements and overall appearance should provide a reasonable indication of its overall well-being.

External parasites are rare in geckos except for red mites, which occur regularly in some species. These mites are not as detrimental as the infamous snake mite and can be treated quite easily *(see Health & Diseases)*, so their presence is not necessarily a major disadvantage as long as the gecko's general condition still appears good. The main health related condition to look for is the retention of pieces of old unshed skin, particularly around the eye sockets and the toes. If there are patches of milky white skin, ask questions about the animal's husbandry. It could be simply due to a low level of ambient humidity in the enclosure, which can easily be addressed. If in doubt as to its cause then it may be best to avoid that lizard, particularly if it also appears low in body condition.

Handling Geckos

As mentioned above geckos are not lizards that take particularly well to handling. If you want a lizard that will sit on your hand you should look at a bearded dragon. Geckos are fast, agile climbers that can be quite delicate if treated roughly. Keep handling to an absolute minimum. If the lizard is to be picked up, always attempt to grasp it around the body directly behind the head and restrain it firmly without squeezing. Some species react better than others and may even become somewhat conditioned to this, although most never really enjoy the manipulation and will continuously struggle and seek escape. Never catch a gecko by its tail as these are readily dropped if the animal is stressed and although this is not necessarily detrimental to the lizard it is an indication of excessive stress and a gecko with no tail is a sad and sorry sight. However, with proper care and feeding, the tail will regrow over a period of several months. Keep fingers away from the mouth of the larger species such as giant cave geckos *(Pseudothecadactylus lindneri* and *P.cavaticus)*, centralian knob-tailed geckos *(Nephrurus amyae)* and ring-tailed geckos *(Cyrtodactylus louisiadensis)* as they have very strong jaws and can inflict painful bites.

Handling legless lizards is even more problematic. Like geckos they are not comfortable with being lifted and restrained but the problem is exacerbated by their reaction and their morphology. When first lifted the legless lizard will often remain still and apparently calm for a few seconds. This is usually quickly followed by a frantic writhing of the body and tail and unless the animal is

Photo by Kelvin Marshall.

released, may result in the loss of the tail, which as described above can be a traumatic experience for reptile and keeper alike. Do not try and use a snake hook to handle legless lizards. They will feel vulnerable and uncomfortable with the small point of support and, if used in the same manner as with snakes by also holding the tail tip, it could be a recipe for disaster.

Photo by Kelvin Marshall.

Examples of appropriate handling techniques for knob-tailed geckos. Faster more agile species may require additional restraint during handling.

Photo by Darren Green.

Keeping Geckos

HOUSING

Enclosure Construction

Geckos are generally small and relatively inactive lizards and are thus tolerant of small enclosure sizes. In general, a pair or a trio of one male and two female geckos may be maintained in an enclosure measuring 600mm wide x 400mm high x 300mm deep, although the smaller species and juveniles will certainly thrive in less than this. The orientation of the enclosure is dependent on the ecology of the species concerned. Primarily ground-dwelling species such as knob-tailed geckos and some of the *Diplodactylus* species need little in the way of climbing height and the smaller species may be adequately maintained in a cage less than 200mm high. At the other extreme, leaf-tails, giant cave geckos and other species that inhabit trees, cliffs or escarpments prefer extra vertical height so an enclosure measuring 400mm wide x 600mm high or even taller would be more appropriate. The depth of the enclosure, i.e. from front to back can stay the same for all species; a narrow cage aids easy care and cleaning by making all areas easily accessible. Legless lizards are generally elongate animals and thus require relatively longer housing. Some species also enjoy climbing amongst vegetation, so some degree of vertical height is also beneficial. Depending on the species, an enclosure of 800x400x400mm is generally adequate for one or two specimens.

A wide range of materials are suitable for housing, all having advantages and disadvantages. The final position of the cage will somewhat dictate which materials will be appropriate for the job. If the enclosure is to be kept outside where it will be exposed to the weather, then waterproof or at least weather resistant materials will be necessary. Even if the unit is maintained under a roofline outside, some timbers such as MDF or particleboard should not be used without adequate protection. Treated timbers may be used but only after they have been allowed to weather for several weeks before the lizards are introduced, to ensure that any toxic chemicals have leached out. Glass, perspex, galvanised steel and mesh and aluminium are all durable materials suitable for exterior cages.

In interior cages, glass is easy to keep clean and provides excellent viewing of the inhabitants but is unforgiving and may be difficult to work with if you are building your own cages. Perspex is also useful but discolours and scratches with time and wear. Timber provides a more natural appearance but must be regularly maintained, especially when species requiring high humidity are housed. Likewise metal may also suffer from moisture within the enclosure.

Personally, a combination of timber or glass is preferred for adult geckos. A basic timber box with a hinged glass door at the front is an excellent start for many species. Use plywood rather than particleboard or MDF as the latter are more prone to deterioration from humidity. Two or three coats of a polyurethane

Example of a bank of enclosures suitable for a range of gecko species including terrestrial and climbing types.

sealer will aid the longevity of the timber, though make sure the construction has been well ventilated before introducing any reptiles. Where a moist substrate is important, glass should be used, as even protected timber will begin to suffer in time under these conditions.

Ventilation is one aspect of lizard housing that is regularly neglected. A consistent flow of fresh air through the enclosure is absolutely essential to maintain the health of the animals within. Stale, humid environments are breeding grounds for the proliferation of noxious bacteria and fungi. By designing an enclosure that has vents on one side at the bottom and on the other side at the top, a passive flow of warmer air leaving the top vents is immediately replaced by fresh air drawn into the bottom. Note that such a situation is far removed from a draught where the flow is quite vigorous and potentially hazardous to the inhabitants. Where it is not possible to install vent holes in the sides of a container e.g. a glass aquarium, it is important that the majority of the top is ventilated. Alternatively, the tank can be taken to a glass merchant who can cut or drill holes in the glass, suitably sized and placed to promote good ventilation.

A long enclosure such as this is required to house larger legless lizards such as the common scaly-foot *(Pygopus lepidopodus).*

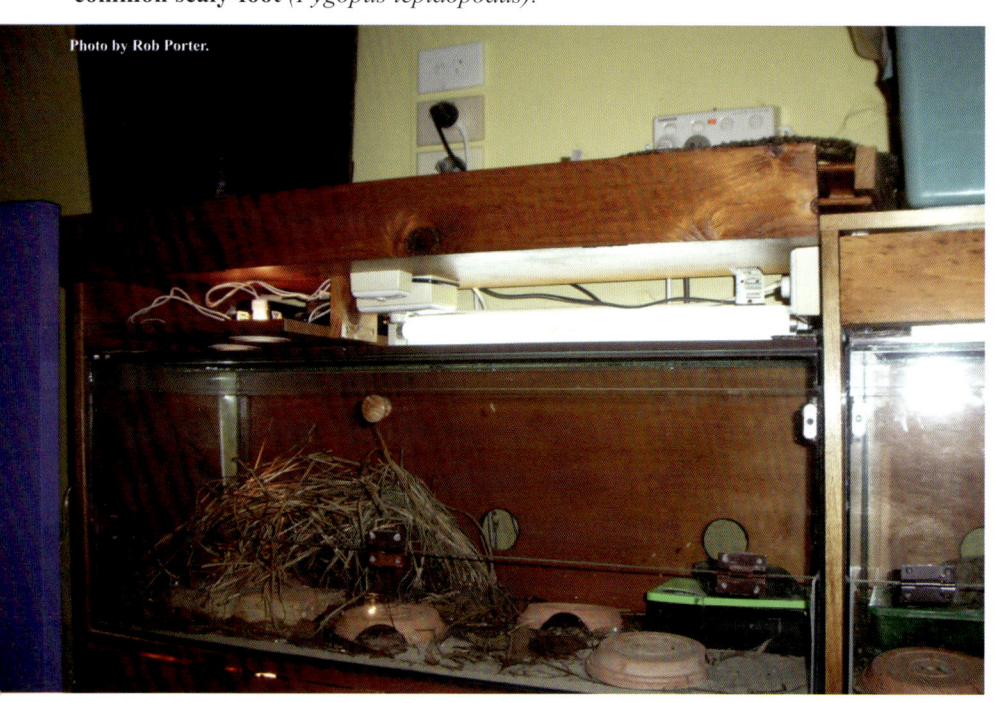

Photo by Rob Porter.

All vents should be covered with strong mesh of a suitable size to contain both the reptiles and their live food items. Metal fly screen mesh is ideal as it is strong and durable and can be used for both interior and exterior enclosures. Do not use nylon mesh, as crickets will quite happily incorporate this as part of their captive diet. For extra security, a combination of fly screen and 12mm weldmesh is even better, particularly if they are mounted on each side of the cage wall material to provide a buffer zone between. This is an ideal way to

---

Although requiring a little extra maintenance, gecko enclosures can be set up to be an attractive feature in the house, this is a commercially produced enclosure available at good reptile petshops.

Photo by Jason Goulding.

keep inquisitive dog or cat claws away from the lizards. Pegboard should not be used unless covered with insect mesh, as the holes are too large to confine crickets and cockroaches. Geckos and legless lizards do not usually tend to rub their snouts along mesh vents, a behaviour commonly observed in dragons and monitors, so the vents can be positioned within the enclosure to take full advantage efficient airflow.

Make provision for good access to all areas of the enclosure with a large opening door at the front or top. This will make maintenance and cleaning an easier task, which will therefore be carried out more regularly than if the cage is awkward to reach. Once settled geckos rarely try to escape from enclosures when the door is opened for cleaning or feeding. If you are concerned about the flighty nature of your gecko, then install an additional smaller door for spot cleaning and introducing food while also keeping the larger door for major cleaning.

Most geckos are best kept indoors where it is easiest to control the enclosure's environment. Make sure that the cage is not placed where it will receive hot afternoon sun through an adjacent window or directly in front of a heater or air-conditioning unit. If it is planned to keep the lizards outside then plan the enclosure's location very carefully. Beneath a veranda or other roofline should be suitable as long as elevated temperatures are not experienced. If excessively hot days are expected, move the enclosure somewhere cooler or spray the inside regularly throughout the day with water to lower temperatures by evaporation. If the site is out in the open it should be protected from direct sunlight and harsh winds by overhead trees. A little dappled sun is fine, but it is heat that will cause the death of the animals more readily than cold. Most geckos and legless lizards, except the tropical species, will handle cold winter temperatures as long as they have somewhere sheltered and dry to hide. This can be provided by covering part of the top of the enclosure with a sheet of corrugated iron or plywood to keep part of the inside sheltered from the rain.

Housing for juvenile geckos requires a somewhat different tack. Once these tiny lizards begin to eat, it is essential that food is readily accessible and easy to find. Consequently, a small enclosure or container is used to house geckos for the first six to eight months of their lives. Plastic food containers are ideal and may be obtained cheaply from variety stores. A section of the lid can be replaced by fly screen mesh melted into the plastic using a soldering iron. Use tall, skinny containers for climbing species and squat, wide containers for terrestrial species. Clear lunchbox-sized containers are great for one or two hatchling geckos of most species for the first six to eight months of their lives.

Heat & Light

Having activity periods during the cool of the night also means that geckos do not require high temperatures for normal metabolic functions. Some heat is still necessary to enable the lizards to undergo many of the processes of living such as hunting, digestion and breeding. Many species are content at temperatures in the mid to high 20ºC's, though this is dependent on the distribution of the species concerned. All enclosures should provide the inhabitants with a temperature gradient; a gradual change in temperature from the warmest part of the cage containing the heat source to the coolest. Tropical species naturally prefer higher ambient temperatures than their southern counterparts and a temperature gradient of mid-twenties to 36ºC at the heat source is ideal during the warmer months for these northern species. For temperate species, low twenties to a maximum of 30ºC is suitable. During winter temperatures may be dropped substantially and this is essential in some species to ensure successful reproduction. Most species except the true 'tropicals', which will require a minimum limit of around 16ºC, can go without any enclosure heating whatsoever over winter. If you prefer to offer some heat then use a gentle and discreet daytime heat source in one corner such as a 25-watt bulb. Others will tolerate night temperatures down to single figures with no ill effects. It is a good idea to invest in an inexpensive digital thermometer to monitor temperatures at both ends of the temperature gradient. Many models come with a probe to enable two separate areas to be measured. A model with a minimum/maximum memory will also illustrate what the gradient runs at during the coolest and hottest parts of the day.

This reduced heat requirement means that low output heat sources are ideal for geckos. This can take the form either of a heat pad on the floor or attached to a wall of the enclosure or a low wattage light bulb. If the latter is used, it should be a coloured bulb, preferably blue or red, as the lizards are less sensitive to these colours. These lights may be left on 24 hours a day during the active time of the year and therefore nocturnal activity periods will not be disrupted while adequate viewing light permits observation of the captives to assess their health and well being and to enjoy watching and learning about their behaviour. Low wattage ceramic heat emitters can also be useful, as they produce no light. However, they must be protected by a barrier to ensure the lizards do not come in contact with the heater. It is a good idea to run these heat sources through a rheostat or dimmer switch to provide finer control over the temperatures at the heat source. Many products produce heat that is too fierce for geckos and a more gentle output is needed. A thermostat is generally not necessary because it supplies an 'all or nothing' service unless one of the expensive pulse proportional units is used. However, thermostats can provide a safety backup for those really hot days, turning the heating off completely if temperatures exceed a particular maximum level.

Whatever heat source is provided, it is imperative that its position creates a temperature gradient within the enclosure that is useable by the inhabitants. Reptiles have an innate sense that tells them what temperature their body should be maintained at to enable all bodily processes to function efficiently. Consequently, the lizards will choose what area of the gradient is most appropriate at a given time. A lizard that has recently eaten will often seek a warmer area to enhance its digestive processes. All the keeper has to do is provide a temperature span that encompasses the required range of the species concerned. For ground dwelling species their useable environment is somewhat two-dimensional so a heat source at one end of the enclosure is adequate. Geckos that are adept at climbing experience a three-dimensional space so a bulb in the top corner of the cage will make full use of the available area.

As nocturnal animals, heating and lighting are far less important for geckos than for their sun-loving relatives such as dragons, goannas and even legless lizards. Ultraviolet light seems to play little or no role in the health of Australian geckos probably because their night activity has forced them to utilize other methods of calcium absorption from the gut. This process is normally controlled by Vitamin D3, which is synthesised by exposure to these ultraviolet light wavelengths. Having said this, there is still some evidence that certain species will partially emerge in sunlight, perhaps to assist with thermoregulation, and by doing so may have developed a requirement for the associated light wavelengths. Species such as the spiny-tailed gecko *(Stophurus ciliaris)* and the golden-tailed gecko *(S.taenicauda)* may remain exposed to filtered sunlight throughout the day as they may rest on exposed tree branches. If providing a daytime light source for your geckos, and it is important that they do receive a photoperiod (i.e. defined light and dark periods over 24 hours) of some form be it natural or artificial; an ultraviolet-producing light would certainly not do any harm. However, there is currently no evidence to suggest that any Australian geckos will suffer in captivity because there is no access to ultraviolet light.

The legless lizards on the other hand are often more diurnal in their activity cycles. It is therefore probable that their requirements for both heat and ultraviolet light is much greater than that of their true gecko relatives. Consequently, both an ultraviolet source and a basking light should be supplied for these lizards. The temperature beneath the main heat source will again vary slightly depending on the natural distribution of the species housed. As a guide, a temperature gradient of 24 to 38°C should be adequate for most pygopod species. It is a good idea to watch the behaviour of your reptiles in relation to the heat source regardless of type to provide clues as to how appropriate the current temperature regime is. If the inhabitants are spending all day beneath the heat source and rarely venturing away from it then it is likely that the heat needs to be increased. The reverse is also true if the heat is too strong. Ideally, the lizards should be seen to bask for short periods then move away to forage, interact with cage mates, etc. then return to bask again, a behaviour known as 'shuttling'. This pattern is then repeated through the day, though it will be somewhat dependent on the time of year.

Moisture & Humidity

Another important environmental parameter is moisture. Many species of gecko require high humidity levels at certain times of the year without which skin-shedding problems and dehydration may develop and ultimately the animal may die. The best method to provide humidity within the captive environment is by spraying the enclosure with water, preferably as a fine mist. A cheap hand-trigger spray bottle will suffice for one or two enclosures but if a larger collection is maintained it is worth investing in a five or six litre pump-action garden sprayer. Generously spray all surfaces within the enclosure but do not saturate the substrate. Some species may enjoy being sprayed directly with tepid water and will start licking droplets immediately. Others, such as the broad-tailed geckos *(Phyllurus platurus)* for example, seem to detest being sprayed and will react quite dramatically by squeaking loudly and waving the tail from side to side. The regularity of this spraying is dependent on the species concerned, the ambient environment around the enclosure and the time of the year. A cage maintained in an artificially heated or cooled room will dry out very quickly despite liberal spraying. For most species a twice weekly application is sufficient, reduced to once every week or so in winter. It is very important that the enclosure is not over-watered, especially if there is insufficient ventilation. The majority of the cage should dry out completely within a few hours of spraying, the exception being those species requiring an area of permanent moisture as described below. Health problems will quickly materialize in many species if damp conditions are maintained.

As you would expect, rainforest species, such as leaf-tailed geckos, require a generally higher humidity than most other gecko species and it is a good idea to also supply a permanently moist area with in the enclosure. This can take the form of a plastic sandwich box container around 250x150x80mm high with the lid removed and filled with a fine sand/coco peat mix. Water is added to the substrate as required to maintain its moisture level. This container will function in two ways. Firstly, the water will slowly evaporate into the enclosure to maintain the humidity and secondly it will serve as an egg-laying site for the female leaf-tailed geckos. It is also essential that some desert geckos, e.g. the smooth skinned species of knob-tailed gecko, be provided with one part of their enclosure that is permanently moist and humid. This can be simply accomplished by maintaining a hiding place beneath which the substrate is watered regularly and is never allowed to dry out completely. Alternatively, a small plastic container with an access hole in one side can be installed in the enclosure and filled with a moist material such as fine sand or sphagnum moss. Make sure that the access is unhindered and check the substrate regularly as it will still dry out even in a semi-enclosed container. Again, both of these items will also provide the perfect egg-laying site so the new eggs will not dehydrate if they are not located immediately.

Geckos usually shed their skin in one piece at a frequency that is dependent on age, growth, season, etc. The starred knob-tailed gecko *(Nephrurus stellatus)* **here exhibits a ghostly white colour immediately before it starts to peel away the old skin.**

Photo by Troy Webb.

Photo by Troy Webb.

Australian Reptile Keeper

A small bowl of water should also be provided at all times for adult lizards. Many species appear reluctant to drink directly from such a receptacle. However, it is always safer to supply a bowl in case it is required. Be careful not to wet the sand around the water bowl when filling as some females may try to lay their eggs around or even in the water bowl where they will either dehydrate or drown. It is best not to provide a water bowl for hatchling geckos as they may fall in and drown or their small insect prey will all end up dead in the bowl fouling the water and leaving no food for the youngsters. Alternatively, a small container filled with coarse aquarium gravel and topped up with water will provide access without the risks. A clean piece of sponge will also serve the same purpose but it must be discarded regularly before it becomes soiled. Good hygiene protocols should always be followed and water bowls need to be removed, cleaned and sterilized once a week *(see Hygiene & Health)*.

Substrate

Cage substrate is more important for some species than others. Simple sheets of paper will suffice for many, especially arboreal geckos that rarely come down to the ground. Paper has also been used successfully by some keepers for terrestrial species that do not burrow such as the rough knob-tailed geckos, *(Nephrurus asper* and *Nephrurus amyae)*. At the other end of the scale, the smooth skinned species of knob-tailed geckos such as *N.laevissimus* and some legless lizard species must have a substrate in which to dig, often to a substantial depth. Many will construct their own burrow systems carrying out major earthworks until satisfied with their excavations. Digging usually occurs in a moist medium that does not readily collapse and ideally retains its shape and structure when dry. Fine sand is a personal preference for most species. If a semi-washed sand is used the finest particles will ensure the substrate will bind together when dry and burrows won't collapse, yet the inhabitants will not become permanently coated in the dust that is associated with unwashed media. Do not use coarse sand or gravel as these may be ingested and cause intestinal impactions. Washed beach sand, red desert sand or plasterer's sand are all composed of fine particles that will pass easily through the lizard's digestive tract if swallowed. Some commercial products are calcium-based sands, which can be readily ingested and will breakdown in the acidic environment of the lizard's stomach providing useful metabolic calcium. Other substrates that are suitable include palm or coco peat, fine garden loam or leaf litter. These may be more appropriate for species that require permanently high humidity as the peat has greater water holding capacity than sand. Mixtures of these products and fine sand may also be suitable. Potting mixes should be avoided as many become dusty and water repellent when dry and contain large, potentially dangerous particles of grit, perlite and/or fertilisers.

Substrate depth is critical for burrowing species of gecko or legless lizard or if egg laying is going to occur within the enclosure. For other circumstances a 10mm depth of substrate is sufficient merely to provide a covering for the enclosure floor. Depending on the species concerned, burrowers may need 100mm or more of substrate in which to dig. With the appropriate substrate, burrow cave-ins are not usually a problem and, if they do occur, rarely cause any harm to the inhabitants. In fact some of the main burrowers, such as the smooth knob-tail species can sometimes be found deep under the substrate without any evident means of escape. This may be an example of the geckos plugging the entrance to the burrow with substrate, a natural defensive behaviour to keep predators out.

A smooth knob-tailed gecko *(Nephrurus laevissimus)* **in its home site. Note the moist substrate on one side.**

Other methods may be employed to provide an egg-laying site without filling the entire enclosure with a deep and heavy substrate. A late term gravid female may be removed from her normal housing and placed temporarily in a smaller enclosure with sufficient depth of substrate for egg laying after which the gecko can be returned to her normal cage. Alternatively, a plastic container of suitable depth and filled with moist substrate can be introduced to the enclosure at the appropriate time. This technique has been successful with leaf-tailed geckos but may not be as suitable for terrestrial species such as knob-tails. For other species, which do not excavate large holes for their eggs (e.g. *Oedura* spp., *Diplodactylus* spp.) a lidded container of moist sphagnum moss with a small access hole cut in the top or side will also serve as a good oviposition site. This has the other advantage of allowing easy inspection to ascertain if and when eggs are laid without major disturbance to the enclosure.

Ideally all the enclosure substrate should be removed and replaced every three to six months. The regularity of this change will depend on the stocking levels in the enclosure and the time of year. A large enclosure holding only one or two lizards may only need a substrate change every ten to twelve months. Likewise periods of inactivity and non-feeding such as where geckos are cooled over winter will mean soiling of the substrate will be much reduced. Do not attempt to wash old substrate and return it to the cage. Unless this is carried out in such a way that the medium is completely sterilized it can lead to the rapid build-up of bacteria or other pathogens within the confines of the enclosure. Use a cheap readily available substrate that can be discarded and replaced with fresh material at each major clean.

Enclosure Furnishings

Apart from the substrate, other enclosure furnishings can be very much a matter of choice as to whether a purely natural set-up is preferred to a primarily artificial design. Natural materials such as rocks, logs, branches, etc. are aesthetically pleasing but are not essential to the well being of the lizards and may be difficult to keep clean. If rocks are used, always make sure they are stable and cannot be undermined by digging. Ideally they should be placed directly on the floor of the enclosure with substrate filled in around them to inhibit lizards trying to gain access beneath. Natural stone or artificial equivalents such as bricks can serve an important role in thermoregulation for captive reptiles as they retain heat during the night and provide an important heat source once the lights are out, particularly for nocturnal species like geckos. However, wherever possible lighter materials such as timber, bark, plastic, etc. should be used in preference to heavy and potentially dangerous objects.

Geckos are usually secretive animals, especially during the day, and need to have somewhere secure to hide. Some species prefer to secrete themselves in narrow gaps between rocks or tree bark. This can be simulated with strips of thick rigid tree bark stacked either vertically or horizontally. Bark is much lighter than rock and safer to use in this situation. Alternatively, pieces of plywood can be stacked together with spacers attached to maintain a suitably sized gap between the pieces. For most other species a refuge on the floor of the cage can provide a good home site. One of the simplest techniques is to place several upturned plastic plant pot saucers around the enclosure. Each saucer has had a small access notch cut into one side and is pushed into the substrate allowing the lizards to wedge themselves between the top of the saucer and the substrate. Reptiles feel much more secure when hiding if they have something in contact with their ventral and dorsal surfaces. Small plastic containers with lids can also serve as hide boxes. Simply cut an access hole in the top or side and fill with dry sphagnum moss or leaf litter. There is also a huge range of products available from pet shops and web sites, useful as onclosure furnishings. Some of these are purely aesthetic while others also serve a practical purpose. Small

A Centralian knob-tailed gecko *(Nephrurus amyae)* **using a broken coconut shell for shelter.**

Photo by Darren Green.

Australian Reptile Keeper

terracotta refuges with water reservoirs on top provide excellent humid home sites where required. The water slowly soaks into the porous terracotta and evaporates, producing a cool humid microenvironment within. Another useful product is a two-piece refuge where one section goes inside the enclosure and the other side attaches to the outside of the glass. A refuge area inside provides a secure homesite for the cage inhabitants where they can be observed through the glass by removing the outer piece without unduly disturbing the animals inside.

Tree branches and bark strips placed diagonally through the enclosure and wedged firmly into place provide excellent access to all areas of the enclosure for arboreal geckos. They will need to be replaced as they become soiled, as it is not really practical to try to clean these objects efficiently. Check all items thoroughly before removal to ensure no geckos are hidden within. Artificial

An example of a tall gecko enclosure suitable for climbing species such as leaf-tailed geckos and cave geckos.

Photo by Kelvin Marshall.

plants may also be incorporated for a more 'wild' look. Live plants are not recommended as they do not thrive under the low light intensity required by geckos, and their moist substrate can lead to a proliferation of detrimental micro-organisms, such as bacteria and fungi that may ultimately lead to health problems with the lizards. Many legless lizards are associated with grassland or tussock vegetation in the wild and in captivity individuals often feel most secure when they are able to curl up in a clump of grass. A fresh grass tussock with most of the soil teased from its roots can be placed in the enclosure and removed at a later date when it starts to deteriorate. Live grass plants in pots can also be installed but these rarely last long in such an artificial set up.

NUTRITION

Food

All geckos and most legless lizards are generalized invertebrate eaters. In other words they will readily take a wide variety of insect and arthropod types as they locate them in the wild, although there are some exceptions to this rule. Some geckos, such as the Centralian knob-tailed gecko, have developed a taste for consuming other lizards, especially smaller geckos and will eagerly run their cousins down for the sake of a nutritious meal. In captivity, however, they will thrive on a diet solely of insects. On the other hand, Burton's legless lizard *(Lialis burtonis)* will only eat other lizards, particularly skinks, and has rarely been successfully weaned onto any other dietary item. This makes their care somewhat problematic and anyone wishing to maintain this species should check with their relevant state legislation with regards which lizard species can legally be used as food items. At the other extreme, some of the smaller desert species of gecko have become termite specialists and will sometimes refuse other insect types in captivity, although most species can be enticed to eat similarly sized crickets with some perseverance.

In general though most species are not fussy about their diet and standard commercially available insects such as crickets and cockroaches are ideal. Some keepers like to vary this diet with wild caught insects such as grasshoppers, beetles, spiders, etc. This additional variety can be beneficial provided there is no risk of contamination of these wild-caught prey items by pesticides or other chemicals. While variety may offer nutritional benefits, many species of gecko and pygopod have been maintained in captivity on a sole diet of commercially available insects for many generations without any problems.

Crickets are probably the food of choice as they are easy to handle and are very palatable to most species. Cockroaches may also be offered but are occasionally rejected by some geckos and their speed and secretive habits may enable them to escape detection within the enclosure and become a nuisance, living and possibly reproducing amongst the cages furnishings. While this may sound advantageous the reptilian inhabitants may become stressed due to the sheer number of these insects, plus their nutritional value declines rapidly as there is little if any food available to the cockroaches within the lizard's enclosure. Consequently, it is recommended that if cockroaches are used as food items they should be offered by forceps to the individual lizards, which will quickly learn to accept food provided in this fashion. Forceps-feeding does have other advantages as well. By directly providing food items to each inhabitant there are no concerns that some lizards are missing out on their quota and the keeper can be confident that the insects are fresh, nutritious and recently dusted if mineral/vitamin supplementation is added *(see Supplementation)*. This technique is especially useful for some of the more aggressive feeding legless lizards such as the common scaly-foot *(Pygopus lepidopodus)*. Care is also required to make sure the lizards do not inadvertently bite the metal forceps during this process as this may cause jaw or tooth damage.

Liberating a large number of insects into the enclosure for the lizards to find later is not usually a good idea for the reasons stated above, however, if a number of prey insects are introduced into the enclosure to provide food for an extended period, e.g. during holiday periods, it is a good idea to add a piece of carrot, pumpkin, or other solid vegetable plus a few dry dog pellets as temporary food sources for the insects. Where several individuals are housed together, feeding time can be somewhat hazardous particularly when two individuals zero in on the same prey item. As with snakes, the feeding instinct is so strong that serious bites and injuries may be inflicted on each other. Skilful use of forceps to keep supplying each lizard with food items can nullify this problem.

Forcep feeding a cricket to a common scaly-foot *(Pygopus lepidopodus)*.

Mealworms are another readily available food source for captive geckos. Some species will refuse these beetle larvae but others will accept them with gusto. They should only be used very sparingly however, as their exoskeleton is quite indigestible and can cause blockages in the digestive tract. They are a nutritious supplement to the gecko's main insect diet but try not to offer them more than a couple of times a month and never more than one or two per lizard. Ideally choose the white mealworms, which have recently shed their skin and are thus more easily digested.

Some species are also partial to soft fruit of various types. A few Australian geckos will lick items such as pear, banana, paw paw, etc. but rarely relish it the way some overseas species do. On the other hand some of the legless lizards, particularly the scaly-foot species will eagerly track down a small dish of fruit placed in their enclosure and, after a few investigative licks, will promptly consume even quite large pieces of sweet fruit. Fruit is also an excellent way of adding supplements to the diet of these lizards. Pinky and fuzzy mice are sometimes used as food items for the larger geckos. These are not recommended except as a very occasional treat or in an attempt to quickly improve the overall condition of a lizard that is already low in weight for some known and reversible reason. However, invertebrates are far better at achieving this via a more natural route even if it does require more work and these should be provided wherever possible in preference to rodents.

The size of the prey offered to geckos should be no larger than around two-thirds the width of the head. This is somewhat dependent on the type of prey item; a soft-bodied insect such as a cricket or caterpillar will be easier to consume than a hard-shelled beetle for example. Some hatchling geckos are very small and consequently require appropriately sized prey. For the smallest species pinhead crickets or fruit flies may be required. The latter can be cultured but if only small quantities are needed collection from the wild by placing a jar baited with over ripe fruit in a shady spot in the garden may be a better option. It is also important that the quantity of food provided is controlled. Many geckos especially juveniles, become extremely stressed if large numbers of live insects are added to the enclosure at once. They will often refuse to eat and it is possible that some omnivorous insects such as crickets may even turn the tables on the tiny lizards and overpower them in numbers. Only supply enough food for the enclosure residents to consume within a few hours, as surplus prey items will rapidly lose their nutritional value and supplement dusting.

Supplementation

Multivitamin and mineral supplementation is as important for geckos and legless lizards as it is for other lizard groups. A good quality calcium and reptile multivitamin product, which contains vitamin D3, should normally be dusted onto food insects once per week, twice for gravid females and growing baby geckos. At other feeds straight calcium powder should be used without additional multivitamins. Insects can be dusted by placing them in a plastic bag or jar and liberally shaking the appropriate powder over them. A regular shake of the container during the feeding exercise should ensure that all food items have some powder attached to their bodies, although it may be necessary to apply more powder during extended feeding programs. Some insects are better than others at retaining the powder. Young crickets before they have grown their shiny wing covers maintain a full coating for a reasonable time, whereas the glossy exoskeleton of cockroaches and mealworms quickly lose any supplement powder.

It is a good idea to provide a dish of calcium carbonate powder within the enclosure for those species that produce calcareous-shelled eggs such as ring-tailed geckos *(Cyrtodactylus louisiadensis)*. This should be straight calcium and should not have any additional multivitamins included. A small dish such as a bottle top is sufficient as their demands are relatively low. Replace the calcium as it is consumed or becomes soiled.

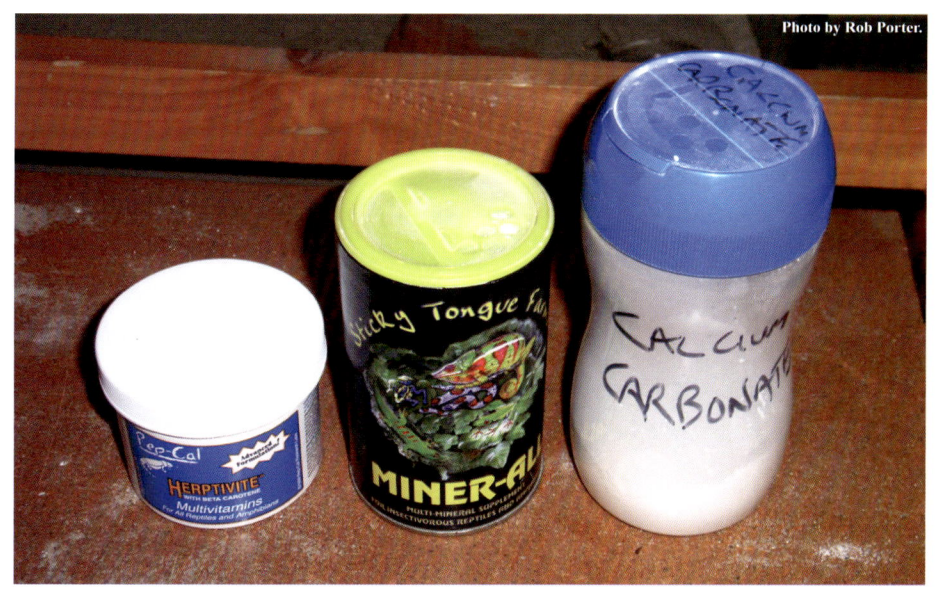

Examples of mineral and multivitamin supplements suitable for geckos and legless lizards.

BREEDING

Most gecko species breed readily in captivity if given adequate housing, mates and the appropriate environment. The same does not appear to be the case for legless lizards. With the exception of the striped legless lizard *(Delma impar)*, which has been successfully bred on a number of occasions, there is little if any evidence of true captive breeding in other species. This may be partly an indication of the rarity of these lizards in captivity. However, certain species, such as the common scaly foot *(Pygopus lepidopodus)*, have been maintained frequently and successfully over the years, yet there are no records of the species ever having been mated and bred in captivity. Despite their close relationship to the geckos, there is clearly much we still need to learn about reproduction in this group.

Seasonal environmental fluctuations are important for some and critical in other species of gecko to ensure regular breeding success. Parameters such as day length, temperature and perhaps humidity or barometric pressure are important cues by which geckos synchronize their reproductive cycles. The type of cues used by each species can usually be determined from their distribution. Temperate species are more likely to be dependent on temperature and day length, whereas tropical geckos may use temperature or humidity. Some species have no season dependent breeding and may reproduce all year round. The degree and length of the fluctuation may also be important. The granite belt leaf-tailed gecko *(Saltuarius wyberba)* appears to need winter temperatures below about 10ºC to ensure captive breeding, while other leaf-tail species may only require temperatures of around 15ºC. If you are unsure of the type and degree of environmental cue to provide your geckos, study its range and distribution, this will provide valuable clues as to which may be important for this particular species. If reproduction is unsuccessful keep experimenting by manipulating the environmental parameters each year until breeding is achieved. Good record keeping is essential for this task to ensure unproductive changes are not implemented again in following years.

Sexing Geckos

In the majority of species adult male geckos are easily distinguished from females by the presence of a pair of swellings at the base of the tail. These swellings indicate the presence of paired male sexual organs called hemipenes. These may fluctuate in size depending on the season and be less conspicuous in some species. They are also not usually evident in juveniles until they are at least six to twelve months of age, though this is dependent on the species concerned and the growth rate. Hold the lizards gently and turn them upside down to examine the vent area for the presence of swellings and if possible examine suspected males and females at the same time to highlight the physical differences.

Vent areas of a male (above) and female (below) rough-throated leaf-tailed gecko *(Saltuarius salebrosus)*. The bulges seen in the male are the paired sex organs or hemipenes.

Keeping Geckos

Males of most species tend to be thinner and less robust than females, particularly around the abdominal area. In knob-tailed geckos *(Nephrurus* spp.*)*, males are distinctively smaller in overall body size than the females though the ratio of size difference varies between the species. In some species males posses enlarged spines on each side of the vent, which are either absent or much less developed in the females.

Behaviour can also be used to assist with sexing, though again this is somewhat species dependent. The males of some gecko groups such as velvet geckos *(Oedura* spp.*)* can be very aggressive to other males especially during the breeding season and particularly if a female is also present. Aggression may manifest itself as overt combat, which even if not directly observed can be assumed when male cage mates develop bite scarring around the head and neck and regular squeaking and other vocalizations can be heard emanating from the enclosure. In some species the females may be equally or more aggressive to cage mates than the males like e.g giant cave geckos *(Pseudothecadactylus lindneri* and *P.cavaticus)*. Either way it is essential that these lizards be separated before any physical or stress-related damage occurs.

Photo by Rob Porter.

Mating in smooth knob-tailed geckos *(Nephrurus laevissimus)*. Note the size difference between males and females of this species.

Australian Reptile Keeper

Mating & Egg Laying

In most cases only one male should be maintained in each enclosure, although three or more females will often cohabit with no aggression problems. However, the males of some species, such as some knob-tail *(Nephrurus)* and leaf-tail *(Saltuarius* and *Phyllurus)* geckos, seem to tolerate the presence of other males. Male velvet geckos *(Oedura)* on the other hand can be extremely aggressive and may eventually kill another male. In giant cave geckos *(Pseudothecadactylus lindneri* and *P.cavaticus)* the females appear to be the dominant sex and can be very aggressive towards additional males or other females, so these geckos should only be maintained as compatible pairs.

Mating usually takes place in late winter or early spring. Males will mount the female and often bite the skin on the back of the neck to maintain his position during mating. In most species the actual mating process is fairly short and takes place at night so it is not always observed. Gestation is around four to five weeks with the first eggs of the season being laid in September/October. Additional matings may occur after the first clutch is laid but are not essential for multiple clutching, as many species are able to store sperm for an extended period of time. Specimens maintained indoors with year round artificial heat and light sources may breed at different times of the year due to the lack of

Mating of the connon knob-tailed gecko *(Nephrurus levis occidentalis).*

seasonal cues to synchronise the reproductive systems. Some of the tropical species will often begin egg laying earlier in the season e.g. giant cave geckos, fringe-toed velvet geckos *(Oedura filicipoda)*, while the larger leaf-tailed species are often the last to start producing eggs each season.

Gravid females are usually noticeable by their increased girth and, in many of the smaller species, the late term eggs can be seen either directly through the skin of the underside or as obvious oval shapes if the abdomen is gently palpated. Test digging by female geckos is usually the first sign that egg laying is imminent. These holes are usually dug one or two days before the eggs are laid and in some of the deep substrate egg layers (e.g. the smooth knob-tailed gecko, *Nephrurus laevissimus*) this may involve extensive excavation and upheaval of the entire enclosure. Once the eggs have been laid it is not always obvious where they have been deposited as the female fills in the hole almost exactly as it was beforehand. Leaf-tailed geckos usually leave a telltale mound of substrate directly above the actual deposition site. For other species, careful monitoring of the female is needed until it is evident that a reduction in her mid-body size has occurred. If a container of sphagnum moss or other moist substrate is used as a laying site this merely needs a quick check every couple of days for the presence of newly laid eggs.

Little information is available concerning captive egg laying in legless lizards. In the wild most species tend to deposit their eggs beneath other objects, possibly due to their inability to excavate an egg laying hole. Common scaly-foot *(Pygopus lepidopodus)* will readily use a plastic egg laying box with a small hole cut into the lid, depositing their eggs on the surface of moist substrate inside.

All geckos and legless lizards throughout the world produce two or rarely one egg in each clutch. Pygopod eggs are usually slightly more elongate and bullet shaped compared with gecko eggs, perhaps a necessary adaptation to their slim body dimensions. Most true geckos produce oval shaped eggs although in some species they may be almost round. The size of the eggs of Australian species ranges from around 6mm in length in the smaller species up to 35mm in the Centralian knob-tail *(N.amyae)*.

Most Australian geckos produce parchment-shelled eggs, which are slightly soft to touch and dehydrate very quickly. For this reason, having a permanently moist laying site within the enclosure is essential. With this technique, eggs that are not discovered for a day or so after laying will still be turgid and viable. If eggs are laid out in the open part of the enclosure they will dehydrate within a few hours of laying. Eggs found in this condition should be placed immediately in a humid environment and may recover. However, if the shells have begun to dimple then most are beyond help. There are several types of oviposition sites that can be offered to ensure successful laying and egg retrieval. For species that prefer to bury their eggs such as knob tails and some leaf-tails one part of the enclosure substrate should be watered every few days during the breeding season and never allowed to dry out. The depth of this area may need to be quite substantial for some species e.g. up to 150mm for the smooth knob-tail *(N.laevissimus)* but most will happily lay in 70 to 100mm including the largest leaf-tails. For species that don't dig burrows but still bury their eggs, an alternative is to introduce a plastic container of permanently moist substrate into the enclosure. For geckos that may deposit them in crevices or beneath rocks, etc., a small container of moist sphagnum moss with an access hole on one side or through the lid will suffice. By using a clear or translucent container, checking for the presence of eggs becomes a quick and easy task as most geckos will deposit their eggs on the bottom of the container and they can be easily seen by examining it from underneath. If the female seems unwilling to use a clear container, the sides can be painted or covered in thick tape to encourage her into a dark laying area while still leaving the bottom clear to inspect for the presence of eggs.

The exceptions to these egg laying requirements are the calcareous shelled eggs produced by a few Australian geckos such as the ring-tailed gecko

Late term gravid female Sydney broad-tailed gecko *(Phyllurus platurus)* **with the two eggs clearly visible through the ventral skin.**

Photo by Rob Porter.

(*Cyrtodactylus louisiadensis*) and Bynoe's geckos *(Heteronotia binoei)*. The hard, impervious shells produced by these species ensure no water loss from the interior of the eggs. These species will often lay their eggs on the floor of the enclosure without any attempt to conceal them, though more commonly they are deposited between or beneath cage furnishings. Care must be taken when handling these eggs as despite the shells being harder than the parchment-type eggs, they are also more brittle and very unforgiving if mishandled

Fertile parchment-type eggs are off-white in colour and firm but slightly flexible to the touch. Any eggs that are soft and very pliable are most likely infertile as are any eggs that are yellowish or golden brown in colour. Viability of parchment eggs will become evident within the first two weeks of incubation as infertile eggs will quickly shrivel and discolour further. Infertile calcareous shelled eggs are slightly harder to pick but may be slightly yellowish in some cases. In most cases it is better to err on the side of caution and set up all eggs as if they are fertile.

This plastic container filled with sphagnum moss with a small hole cut in the side is a suitable egg-laying site for small gecko species such as *Diplodactylus, Strophurus* **and** *Oedura* **species.**

Ring Tailed Gecko
(Cytrodactylus louisiadensis)

Rough-throated Leaf-tailed Gecko
(Saltuarius salebrosus)

Centralian Knob-tailed Gecko
(Nephrurus amyae)

Golden-tailed Gecko
(Strophurus taenicauda)

Examples of a range of Australian gecko eggs.

Egg Care

Clutches of eggs should be removed from the enclosure as soon as they are located and transferred to an incubator. It is very important that all reptile eggs are not rotated away from their laying orientation during transferral to the incubating container. Once the embryo has begun its development any change in its position may inhibit development, particularly during the mid and late terms of incubation. As a safeguard, it is a good idea to mark the top of each egg with a marker pen or pencil before removal to ensure the original position is retained. All eggs should be removed from the enclosure for incubation elsewhere. Some eggs may well incubate successfully where they have been laid in the enclosure but any future digging by cage inhabitants may disturb their orientation or even damage the shell, resulting in the death of the embryos within.

For incubation, gecko eggs can be set up in the standard coarse grade vermiculite or perlite at a ratio of around 1:1 by weight with water or by using sphagnum moss. The author has used the latter substrate successfully for many years. This involves a small 100ml plastic container and lid half-filled with moist sphagnum moss. The eggs are then introduced and the container is filled, but not compacted, to within 10mm of the lid with more moss. The moss is in a ratio of approximately nine parts water to one part moss by weight. This may seem wet but the large number of air pockets within the moss ensures sufficient air exchange across the egg membrane while still maintaining 100% humidity. If the container has a tight fitting lid, one or two small holes could be provided in the side of the container to assist with gas exchange across the egg membrane. Alternatively, the container lid can be briefly removed every seven to ten days to allow replacement of the stale air inside. It is important that the appropriate level of moisture is retained through the entire incubation period. To this end it is a good idea to weigh the container complete with the incubation medium, added water and eggs before incubation begins and then every two to three weeks during incubation introducing additional water as required replacing that which has been lost. When carrying out this operation keep disturbance of the eggs to a minimum, use tepid water and never pour water directly onto the eggs.

Geckos may be divided into two groups according to their incubation temperature. By far the majority of species should be incubated at around 28 to 29°C with a range of 27 to 30°C, a range that also applies to the legless lizards. The exceptions are the eggs of leaf-tailed gecko species and the chameleon or carrot-tailed gecko *(Carphodactylus laevis)*, which must have temperatures of 24 to 27°C, otherwise most will fail to hatch. It is not necessary to maintain an absolute constant temperature throughout the incubation period. In fact there is some evidence to suggest that a natural daily fluctuation in temperature is beneficial to the health of the developing lizard eggs including some gecko species.

Incubation of gecko eggs. Each clutch of two eggs is set up in a separate incubation container inside the incubator. Perlite has been used as an incubation medium in this case.

Australian Reptile Keeper

The length of the incubation period at the appropriate temperature range varies enormously. Some of the smaller *Diplodatylus* species such as the helmeted gecko *(D.galeatus)* have been recorded as hatching after a mere 38 day incubation while at the other extreme ring-tailed geckos may take 150 days or more to hatch. Generally a period of between 50 to 70 days will cover most species with the exception of the leaf-tailed geckos and chameleon gecko, which usually go through an 80 to 110 day incubation due mainly to their lower required incubation temperature. Little information is available for the legless lizards but the few records of successful hatching suggest that a slightly longer period is required than for many true geckos, mostly around 60 to 80 days.

Hatchling Care

Upon hatching, the neonates are immediately transferred to a small plastic raising box approximately 200x150x80mm high for terrestrial species and 150x150x200mm high for arboreal species. By using these small nursery cages, the chance of each individual encountering a food item is increased, improving both the survival and growth rates. Depending on the growth of the species involved, the juveniles may stay in these containers for up to ten months until they are large enough to be moved into a full size enclosure. Faster growing species may need to be moved on after six to eight months.

Each container is set up in a relatively simple fashion with fine sand substrate and either horizontal or vertical strips of bark for hiding places or small pot saucers for species such as knob-tailed, thick-tailed and many *Diplodactylus* geckos. Many species except the burrowers can be raised on paper towels to ensure that excellent hygiene is maintained. Most species can be maintained together as clutch mates in one raising box. Some of the knob-tails, however, appear to thrive if raised separately particularly the rough-skinned species *N.amyae* and *N.asper*. Monitor cage mates carefully and if there is any indication that one is lagging behind in its growth rate then separate them into individual boxes immediately. Baby geckos will shed their skin within 24 hours of hatching but may not begin to feed for as much as ten to fourteen days while they continue to absorb the remains of their egg yolk. On the other hand some species such as many *Diplodactylus* and velvet gecko *(Oedura)* hatchlings will start feeding within a few days of emerging from the egg. Cages should be sprayed with water every second day during the warm months and twice a week at other times, as no water bowl is provided initially to ensure neither geckos nor prey items fall in and drown.

Once feeding commences, small insects such as crickets and cockroaches well dusted with calcium/multi-vitamin powder are provided three or four times per week. Do not provide too many insects at each feed; around two or three food items per gecko should be sufficient. Excessive prey in the cage might cause some of the more nervous species to become stressed by the insects crawling on them and refuse to feed. In addition any insects that are not consumed within the first few hours will lose their dusting powder and will rapidly decrease in nutritional value as their stomachs empty. Poor or impeded bone growth can develop if juveniles are not receiving adequate amounts of calcium in their diet. Observe all inhabitants of raising cages closely to ensure they are all receiving their necessary quota of food.

Most species will reach sexual maturity after around two to three years in captivity. Some species may grow at a much more rapid rate and some of the smaller velvet geckos and *Diplodactylus* species will breed in their first year if kept warm and given plenty of food. What effect this rapid growth rate may have on the life expectancy of the lizard remains to be seen. Other species, such as the large leaf-tails are much slower in growth and in some maturity may not be achieved for four or even five years. Geckos are surprisingly long-lived lizards. The author has one female southern spotted velvet gecko *(Oedura tryoni)* that is still producing three clutches a year after 20 years in captivity. There is no reason to believe that some of the larger species could live for 20 years or possibly longer.

Hatchling rough knob-tailed gecko *(Nephrurus asper)* **breaks free from free from the egg in which it has developed for the last two or three months.**

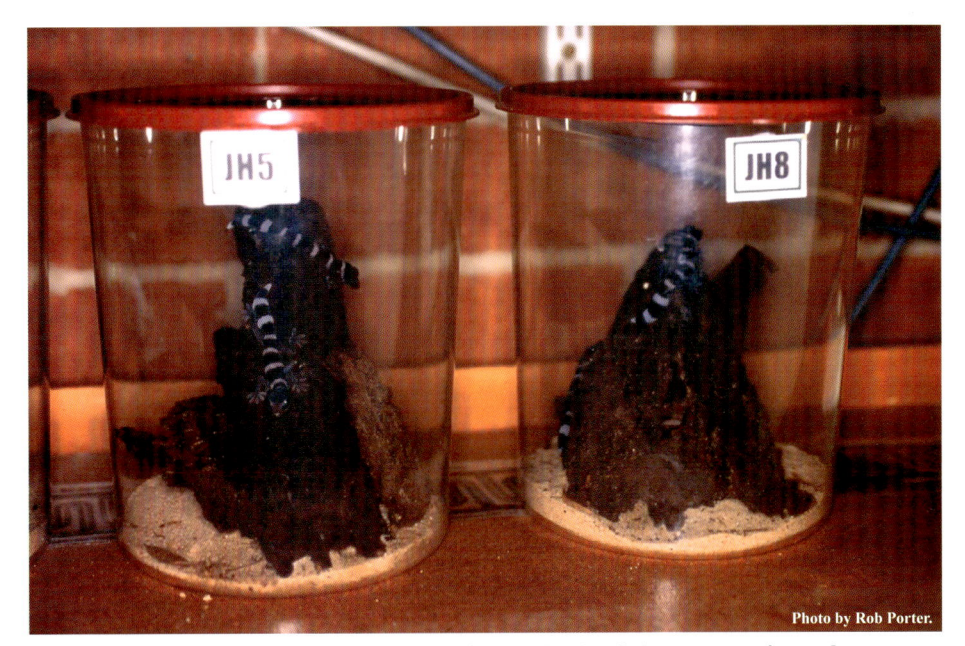

Raising tubs for climbing gecko hatchlings. Each of these containers house two juvenile Kimberley fringe-toed geckos *(Oedura filicipoda)* until they are about 6-8 months of age.

Raising tub for terrestrial gecko hatchlings, such as baby *Nephrurus* and *Diplodactylus*.

Keeping Geckos

Species	Incubation Temperature	Incubation Period Days	Reference
Carphodatylus laevis	24°C	96-113	Personal observation
Cyrtodactylus louisiadensis	28-30°C	130-150	Personal observation
Delma impar	28-31°C	42-47	Banks et al 2000
Diplodactylus conspicillatus	27-30°C	59-65	Brown 2005
Diplodactylus galeatus	27-30°C	46-55	Brown 2005
Diplodatylus tessellatus	28°C	36-45	Laube 2002
Lucasium byrnie	28°C	38-46	Laube 2002a
Lucasium occultus	27-30°C	42	Brown 2005
Lucasium steindachneri	28°C	45-54	Laube 2002
Nephrurus amyae	27-29°C	69-84	Personal observation
Nephrurus asper	27-29°C	68-94	Personal observation
Nephrurus laevissimus	27°C	71-80	Laube 2001
Nephrurus levis	27°C	82-88	Laube 2001
Nephrurus milii	26.5°C	73-82	Laube & Porter 2004
Nephrurus sphyrurus	27-29°C	59-73	Laube & Porter 2004
Oedura castlenaui	28-30°C	58-65	Personal observation
Oedura filicipoda	27-29°C	64-81	Personal observation
Oedura marmorata	28-30°C	60-70	Personal observation
Oedura tryoni	27-30°C	51-58	Personal observation
Phyllurus caudiannulatus	23-27°C	60-85	Personal observation
Phyllurus platurus	22-26°C	66	Personal observation
Pseudothecadactylus cavaticus	28-30°C	70-85	Personal observation
Pseudothecadactylus lindneri	29-33°C	49-74	Sonnemann 1998
Pygopus lepidopodus	28-29°C	65-80	Personal observation
Saltuarius cornutus	24-27°C	66-76	Personal observation
Saltuarius salebrosus	23-27°C	81-99	Personal observation
Saltuarius swaini	22-26°C	72-106	Personal observation
Saltuarius wyberba	23-27°C	67-75	Personal observation
Strophurus ciliaris	27-30°C	50-80	Brown 2006
Strophurus intermedius	27-30°C	59	Brown 2006
Strophurus taenicauda	27-29°C	52-77	Personal observation
Strophurus williamsi	26-30°C	48-56	Laube 2000

HYGIENE & HEALTH

Most species of Australian geckos make hardy captives and are not overly fussy about their captive environment provided basic hygiene and husbandry techniques are followed. Species do differ somewhat in the degree and fashion by which their enclosure becomes soiled, which will have a bearing on how the cage is maintained. Most species produce their waste products as a single pellet containing faecal matter and urates, the latter seen as a white part of the pellet. This group includes the knob-tail and leaf-tail geckos and the *Diplodactylus* species. Spot cleaning for these lizards is a relatively simple task once per week. Extra vigilance is needed around the moist areas of the enclosure or egg-laying containers, as the high humidity is very conducive to the growth of detrimental bacteria and fungi.

By contrast the velvet geckos *(Oedura* spp.*)* and others such as giant cave geckos *(Pseudothecadactylus lindneri* and *Pseudothecadactylus cavaticus)* are extremely messy geckos that produce both pellets and considerable amounts of separate urate deposits. These white, chalky materials are left scattered around the enclosure surfaces and quickly become an eyesore as well as a hygiene issue. Scrubbing the enclosure furnishings and other surfaces with an appropriate disinfectant such as a 5% bleach solution is the only way to remove this soiling and this needs to be carried out regularly, especially in enclosures constructed from porous materials such as timber. Provided this regular spot cleaning is carried out efficiently the only other regular maintenance procedure is to remove and sterilize the water bowl once every 1 to 2 weeks. This can be simply achieved by soaking the bowls in the 5% bleach solution for about 15 minutes.

Ideally, food insects should be offered to the lizards and then removed if uneaten after a few hours, however, this is not always possible especially in a large collection. It is a good idea to check the enclosure thoroughly once a week and remove any live insects that may still be running free. These food items are long past their use as a nutritious food item as they quickly empty their stomachs of food and have to resort to eating lizard droppings while also losing any multi-vitamins or minerals that had been applied by dusting. Any geckos preying on these items will not only obtain little nourishment but may also contribute to their internal parasite load by reinfecting themselves with parasites that the insects have ingested from the lizard's own droppings.

Depending on the number of geckos housed, the entire enclosure should be broken down every three to six months and the substrate replaced with fresh material. The furnishings should be removed and either discarded or cleaned as described above and allowed to dry in direct sunlight before being used again. Natural materials such as branches and bark are absorbent and difficult to clean effectively so are better being replaced. Rocks, plastic pot saucers and other artificial refuges can be soaked in 5% bleach solution for 15 to 20 minutes after having any soiling scrubbed off before immersion. After soaking do not rinse but allow these items to dry naturally and make sure they are completely dry before reintroducing the geckos.

Calcium/D3 Deficiency

Calcium is an essential mineral that plays a major role in bone growth and maintenance as well as other functions such as nerve transmission. Although quantities are obtained through a normal diet it is the presence of another material, Vitamin D3, which enables the circulatory system to acquire the mineral from the digestive tract. This vitamin is also found in the diet of some reptiles such as snakes, that consume vertebrate prey. However, insectivorous reptiles have little or no D3 in their diet and must synthesise it themselves through the exposure of their skin to ultraviolet light of particular wavelengths from unfiltered sunlight. This is how day active species such as legless lizards obtain their Vitamin D3 requirements. Geckos on the other hand are primarily nocturnal and thus rarely if ever receive direct exposure to ultraviolet light. So how do they obtain their calcium? The truth is we don't know for sure how this is achieved. There are various theories such as the animals obtain the benefit of reflected sunlight into their daytime retreats or they bask in sunlight very discreetly by exposing a small portion of their body to sunlight while mostly remaining concealed. It is also possible that these lizards have a reduced need for Vitamin D3 or are able to synthesise it by alternative means. It is known that some geckos in captivity can suffer from calcium deficiency, particularly those that produce calcareous-shelled eggs. This latter group actually store calcium at the back of the throat in sacs that can be readily seen if the gecko opens its mouth or, in extreme case, as swollen areas around the neck region. Calcium deficiency may also occur in species producing parchment-type eggs particularly in juveniles that are growing rapidly. This may manifest itself as a kinked or domed spine. Other symptoms of this deficiency may take the form of thin and apparently flexible bones particularly in the limbs or jaws, often associated with swellings in these areas caused by enlargement of ligaments or tendons to counteract the loss of bone strength or soft-shelled eggs that collapse soon after laying.

Regular supplementation with a calcium powder along with a multivitamin mix that contains D3 will usually provide all of the normal mineral requirements. It is possible that gecko's have a reduced need for this vitamin so it may be advisable to restrict its supplementation to once every seven to ten days, whereas calcium powder can be provided at every feed. The best method of adding these items to the diet is to dust feed insects with the powdered supplements by shaking them in a container or bag immediately before feeding. The powder is quickly lost from the insect bodies so it should be applied regularly during the feeding process. Some geckos may still utilise ultraviolet light to produce Vitamin D3. Species such as spiny-tailed *(Strophurus ciliaris)* and golden-tailed geckos *(S.taenicauda)* will often rest during the day on exposed branches and it is conceivable that this behaviour may have led to a reliance on ultraviolet light. For this reason it may be beneficial to provide an artificial ultraviolet light source for these species in captivity. In fact there is no harm in providing these lights for any captive gecko species, as a photoperiod of some type is essential so why not offer this via an ultraviolet producing light. This along with regular supplementation should ensure calcium and/or Vitamin D3 deficiencies do not occur.

Mites

Wild geckos are often found with infestations of small bright orange mites. Due to their colour they are very conspicuous and are most often found around the eye socket, the ear opening or the base of the limbs. Provided the geckos are relatively healthy these mites rarely cause any major health problems and will only increase to serious numbers if the lizard is stressed or already has some other underlying health issue affecting the immune system. Eradication is relatively straightforward and entails regular applications of vegetable oil dabbed onto the mites with a cotton bud. The oil will block the breathing apparatus of the mites smothering them and they will eventually die and drop off within seven to ten days. Repeated applications may be necessary along with thorough cleaning of the enclosure to stop reinfestation.

Gut Impaction

This is a problem that is often associated with captive lizards but actually occurs quite infrequently. The main cause is the accidental ingestion of large particles particularly from the enclosure substrate causing an obstruction in the digestive tract hindering the passing of food or waste products. Sand is often regarded as the worst offender for this ailment, however, if a fine grade of sand is used the particles are small enough to pass through the gut of the lizard and are eventually excreted without any further problems. Washed beach sand, red desert sand or commercially available plasterer's sand are all suitable as enclosure substrates. Coarser materials such as river sand, gravel, perlite or vermiculite should be avoided at all times.

Symptoms of a potentially serious impaction include loss of appetite, lethargy, constipation and a swollen and eventually distended abdomen. In the early stages it may be possible to administer some type of mild laxative such as one or two drops of vegetable oil in the hope that this may assist with dislodging the obstruction. Drops can be placed on the snout of the lizard and these will usually be quickly licked off. However, if the problem persists and the lizard's condition deteriorates veterinary attention is essential.

Sloughing Problems

True geckos slough their old skin in one piece starting around the lips leaving a perfect ghostly replica of the lizard. In contrast, the pygopod skin is shed in several pieces. Some geckos will eat their old skin immediately after shedding possibly providing an important protein source in times of low food availability. Successful skin shedding in all reptiles is highly dependent on a suitable external environment as well as the general health and condition of the animal concerned. Appropriate environmental conditions particularly humidity will ensure the old outer skin separates from the new skin beneath with ease. If this is not the case, small or large portions of the old skin may be retained particularly around the extremities such as toes and tail tip and around the eye socket. If the environmental problems are not addressed and the skin continues to build up in these areas over several sloughing periods then serious long-term problems may occur. Retained spectacles over the eyes may impair vision and therefore the ability to catch prey while the build up of several layers of old skin around toes can lead to a restriction or loss of blood circulation to these areas and their subsequent loss through necrosis.

The dependency on a humid environment for successful skin shedding varies between gecko species. Many of the leaf-tail group are very susceptible to retained skin, which is a reflection of the moist, humid habitats where these species live. Surprisingly, some of the desert species are also tied to these high moisture areas. The smooth knob-tailed geckos such as *N.levis* and *N.laevissimus* must have a section of their enclosure that is maintained permanently moist and while this site may not always be used, the lizards have an innate sense as to when more moisture is required in their surroundings to permit sloughing. In the natural environment these geckos excavate extensive burrows beneath the hot arid surface, inside which the air is constantly cool and often humid.

Other health issues may exacerbate the incidence of retained skin in captive geckos or pygopods but the most common cause by far is an inappropriate enclosure environment. Address any such occurrence immediately by installing a moist, humid retreat of sufficient size to provide the lizards with full access. This may be a corner where the substrate is regularly watered or a separate container with or without a lid in which a moist medium is maintained. Palm peat, sand, sphagnum moss may all be used but stay clear of vermiculite or perlite as these will adhere to the lizard's skin or they may be accidentally ingested causing gut impactions. While these sites may already be introduced for egg laying purposes during the breeding season, it may be wise to provide them all year round to ensure healthy skin shedding.

Egg binding

It is essential that gravid females be provided with appropriate places to lay their impending clutches. The characteristics of these egg-laying sites will vary somewhat between the species and it is the responsibility of the keeper to assess these needs and ensure the correct site is supplied. Females, which are not comfortable with the conditions in their enclosure, will sometimes retain their eggs rather than lay them somewhere they deem unsuitable. For example a gravid female smooth knob-tailed gecko *(Nephrurus laevissimus)* will not lay her clutch unless there is sufficient depth of moist substrate in which to bury the eggs. If this is not provided the animal will pace the enclosure and continuously excavate what substrate is present expending a great deal of energy and becoming increasingly stressed, not a healthy situation in view of her condition. If this situation is not addressed the female will lose what little condition she has at the end of the gestation period endangering both herself and the future clutch. Eventually, the female's condition will become serious and her life will depend on her ability to deposit the eggs. Extended retention of fully formed eggs may lead to the death of the female without veterinary intervention.

The key to this situation is to ensure the female has at least one and preferably several potential egg laying sites. Offering differing types of laying areas may be useful for new acquisitions. Once her preference has been ascertained all other options may be removed for future clutches. By becoming acquainted with the normal gestation periods for the species concerned the keeper can closely monitor the female's condition and become aware of when her egg laying is overdue. There is a little variation in gestation periods but in most cases eggs should be laid within 28 to 40 days after mating. While it may not always be possible to observe matings and even those that are seen are not necessarily successful, regular monitoring of the condition of gravid females will provide a reasonably accurate indication of when eggs are fully developed.

HELMETED GECKO
(Diplodactylus galeatus)

An exquisitely patterned little gecko from the deserts of central Australia, the helmeted gecko rarely exceeds 75mm in total length. It hides from the intense heat of the day in shallow burrows, often beneath rocks, emerging at night to forage for small insects. They are undemanding captives requiring a small low enclosure around 450x300x200mm high with a fine sand substrate around 40-50mm deep. They will sometimes dig their own shallow burrows but appear content if supplied with a number of hiding spots on the surface. Like most other *Diplodactylus* species, they are eager feeders with healthy appetites for small insects like crickets and cockroaches. Some species, such as the fat-tailed gecko *(D.conspicillatus)* can be more problematic to feed as they are specialised termite feeders and may refuse to take anything else. The helmeted gecko will happily live in pairs or trios of a male and two females with breeding occurring in spring and early summer.

Other related species are sometimes available but their small size and relatively bland colouration make them less popular as captives. The stone gecko *(D.vittatus)*, the sand plain gecko *(Lucasium stenodactylum)* and the box-patterned gecko *(L.steindachneri)* are all maintained in small numbers, while some of the more obscure Western Australian species are rarely maintained. Husbandry for most of these species is as described for the helmeted gecko above and the majority will regularly breed with the appropriate captive environment along with a brief winter cooling period. They are not long-lived geckos when compared to some of the larger species but a life span of 8-10 years can still be expected under the right conditions.

Australian Reptile Keeper

The helmeted gecko's *(Diplodactylus galeatus)* common name refers to the pale marking on top of the head. The attractive blotched pattern continues down the body and tail but is usually absent from specimens with regrown tails (above).

GOLDEN-TAILED GECKO
(Strophurus taenicauda)

An adult golden-tailed gecko is one of the most striking of all the Australian geckos. The simple pale grey background overlaid with an intricate black reticulated pattern is beautifully contrasted by the bright orange tail and eye. In some specimens the orange stripe on the tail extends up the dorsal surface to the mid-body region. The effect of this colour scheme is much more impressive when these lizards are active at night. If threatened these geckos will sometimes open their mouth widely to display a dark blue lining. When resting during the day the whole body takes on a darker colouration and the pattern and markings are quite obscure. Unlike most juvenile geckos, the colour of the juvenile golden-tailed gecko is quite bland when compared with the adults, with a dirty wash over all the skin even the golden tail marking is almost indistinguishable in hatchlings.

Their natural range is quite restricted in size inhabiting the eucalypt and native pine woodlands of inland areas of southern Queensland. Although the size of their range is quite small, the species is extremely common in some parts of their distribution with large numbers occurring under loose tree bark and in cracks and crevices of fence posts and fallen timber. This abundance is reflected in their highly productive reproductive cycles. In captivity, female golden-tails can produce 7-8 clutches of two eggs in a season and, if ample food is provided, young females can reproduce in their first year, although this is not recommended. Eggs will be deposited wherever there is a moist area within the cage, sometimes even in the water bowl. Ideally, the laying site will be a small container of moist sphagnum of other substrate and, when a female is close to laying, all other areas must be kept dry. If the water bowl is removed completely, then the enclosure should be lightly sprayed every second day to provide moisture for the lizards.

Captive care is quite straightforward with an enclosure of 600x300x400mm high sufficing for a male and one or two females. They like to climb so provide small diameter branches for access to all areas of the cage. These will often be used as daytime resting areas, the geckos stretching out with the legs pulled in to mimic the shape of the branch. Juveniles should be offered similar furnishings but of a smaller diameter to suit their size.

Australian Reptile Keeper

Golden-tailed gecko *(Strophurus taenicauda)* **adult (above) juvenile (below).**

SPINY-TAILED GECKO
(Strophurus ciliaris)

Another excellent captive species for inexperienced keepers, the various subspecies of the spiny-tailed gecko are beautifully marked and extremely variable in colour. They are predominantly arboreal in habit so the enclosure should contain a network of twiggy branches that the lizards will use by night when they are active and during the day as resting spots. With a maximum length of around 150 to 160mm, spiny-tails are less robust in body shape than the velvet geckos of similar length. This elongate design enables them to sleep on branches apparently exposed but remaining undetected by blending into their environment. This habit perhaps suggests that ultraviolet light may be important to these geckos with their daily exposure to these rays developing a dependency on the light's ability to produce vitamin D3 when it is absorbed by the skin. Consequently, it may be beneficial to these geckos to provide an artificial source of ultraviolet light although they have been successfully maintained for long periods without this form of lighting.

Like most other geckos, spiny-tails are not fussy about their food requirements with most arthropod prey being accepted without hesitation. They have large mouths for their slim build and can therefore consume surprisingly large food items. If threatened, this species and other related geckos have the ability to squirt a thick sticky fluid with a consistency of golden syrup from small glands in their tail. This act rarely occurs in lizards that are long-term captives although if one of these geckos is disturbed without prior warning a keeper may still be the target of this bizarre protection mechanism. The fluid does not smell or seem to cause any irritation to the skin but may just be of enough shock value to a prospective predator to permit the lizard's escape.

A tall type enclosure is ideal for this species with a basking spot at one end of the gecko's enclosure at around 32-34°C. Spray the enclosure liberally with water every few days allowing it to dry out between sprayings. Like other *Strophurus* species, spiny-tails will produce numerous clutches of eggs over the breeding season, which is usually between September and February. Provide a small container of moist substrate at the cooler end of the enclosure and check regularly. Clutches are usually laid about 3-5 weeks apart. Remove these to an incubator at 28-29°C and they should hatch after 55-65 days.

The spiny-tailed gecko *(Strophurus ciliaris)* is often attractively coloured. The spines are not confined to the tail, there are smaller examples continuing up the body and then larger spines again over the eyes.

Keeping Geckos

THICK-TAILED GECKO
(Underwoodisaurus milii)

The thick-tailed or barking gecko has been a popular captive subject for many years. It is an attractively marked medium-sized gecko with a rusty brown body dotted with pale creamy-yellow spots and a black tail with thin bands of white or pale grey. It occurs over much of the southern third of the country but extends north to Alice Springs in the centre and to southern Queensland in the east. It is often associated with rock outcrops but is also found in open woodland.

Thick-tails are undemanding captives requiring an enclosure of 600x300x300mm in size for a pair. They are totally terrestrial so no climbing furnishings are required. They do like to burrow so a relatively deep substrate of fine sand or similar is needed. One corner of the enclosure should be watered once or twice a week allowing it to almost dry o u t between watering. Extra moisture may be required if old shed skin begins to accumulate around the toes. This species is very intolerant of high temperatures so any heating supplied must be very gentle. A maximum hot spot of around 32°C is ample. Ambient temperatures in excess of 35°C for any length of time can be fatal. A winter cooling period is essential to ensure synchronisation of the reproductive cycles. No heat is required at night and temperatures as low as 8 to 10°C are acceptable increasing to 20 to 22°C during the day.

If the appropriate climate is provided female thick-tails should begin producing their first eggs in August/September after the temperature and the day length have started to increase. Females begin frantic digging a day or so before the eggs are laid. Make sure there is a permanently moist area of substrate or a laying container in the enclosure before this time. After laying, the eggs should be transferred to an incubator set to 27 to 29°C and incubation will take 60 to 70 days at this setting. Females are very productive and can continue to produce seven or eight clutches in a single season. The hatchlings measure 60 to 65mm and will begin feeding after six to seven days. The sexes can be differentiated once the juveniles are six to eight months of age and should start breeding in their second year.

Australian Reptile Keeper

Thick-tailed gecko *(Underwoodisaurus milii)*.

Keeping Geckos

COMMON KNOB-TAILED GECKO
(Nephrurus levis)

This is the most commonly maintained species of knob-tailed gecko and one of the easiest to breed. The species is also known as the three-lined knob-tail because of the markings on the back of the head and neck. It occurs over much of central Australia and is further divided into three subspecies, with *N.l.levis* being the most readily available. Although they inhabit some of the most inhospitable country, they are actually quite delicate creatures that are very reliant on cool, humid burrows to shelter them from the intense desert heat. This requirement is reflected in their captive husbandry where one area of the enclosure is kept permanently moist and has one or two retreats for the cage inhabitants. Failure to provide adequate humidity will eventually manifest itself as shedding problems and ultimately the death of the lizards as they become severely dehydrated. All of the smooth skinned knob-tails (e.g. *N.levis, N.laevissimus, N.stellatus*) enjoy a relatively deep substrate where they can excavate their own burrows. Fine sand around 100 to 120mm deep is ideal. As nocturnal desert-dwellers knob-tails do not require high temperatures. One end of the enclosure should be heated to about 32°C by using a heat pad or a low wattage blue or red bulb, which will not affect the animal's nocturnal habits. Winter temperatures can be considerably lower, even down to single figures without causing any harm.

Common knob-tails will breed readily once they are two to three years of age. One male may be maintained with two to three females who will begin laying their clutches of eggs around September after eight to ten weeks of cooler weather over the winter months. Clutches will continue to be produced right through summer and into autumn about four weeks apart, with as many as nine clutches being recorded over one season. Females that are ready to lay their eggs become quite restless and will often excavate extensively for one or two days before actual egg-laying until they are satisfied with the site. Eggs should be transferred to an incubator immediately after laying to ensure they are not disturbed by the excavations of other cage inhabitants. At 28 to 29°C the eggs will hatch in 65 to 80 days. The perfect little replicas of the adults will shed their skin within 24 hours of hatching and are ready to take on the world pushing themselves up high on their tiny legs and waving their tails from side to side if threatened or surprised. Feeding on small crickets and cockroaches will begin four to five days after hatching.

Australian Reptile Keeper

Adult common knob-tailed gecko *(Nephrurus levis levis)* **(above) and comparison of the sexes (below) with the smaller male on the left.**

Keeping Geckos

CENTRALIAN KNOB-TAILED GECKO
(Nephrurus amyae)

This knob-tail is a member of the second group in this genus; the rough skinned knob-tailed geckos, along with *N.asper* and *N.sheai* and *N.wheeleri*. These are predominantly northern Australian species that are characterized by a rough-textured skin from a covering of small rosettes of tubercules or raised scales. The Centralian knob-tail is one of the largest of the Australian geckos at least in bulk if not in overall length, as the tail is almost non-existent except for the tiny round knob. Its impressive size is reflected in its natural diet, which includes other smaller species of gecko. In captivity a diet exclusively of insects is preferable. Some keepers regularly offer newborn pinky mice but these should be avoided unless the reptile is severely depleted of condition such as post-breeding females when they should be offered very sparingly.

Unlike their smooth-skinned relatives these knob-tails do not dig their own burrows and are happy to hide beneath a suitable object such as a piece of bark or an upturned pot plant saucer. The depth of substrate in the enclosure is therefore not so critical and 20 to 30mm is sufficient to meet their requirements. Their different habits also mean they are less reliant on humid microenvironments, so most of the enclosure should be maintained with a dry substrate. However, it is still important to provide one area of substrate that is moistened once or twice a week as this will aid at sloughing time.

Centralian knob-tails are large geckos requiring an enclosure approximately 800x400x300mm high for a male and one or two females, although it may be possible to keep more than one male in a larger enclosure without any undue aggression. The moist corner mentioned above should be maintained throughout spring and summer, as this will be the spot where the females will deposit their eggs. Alternatively, a plastic container with a side access hole could be provided for this purpose. It should be large enough for an adult gecko to crawl inside and turn around easily and it should be partially filled with moist sand or a sand/palm peat mixture, which will hold the moisture slightly longer than straight sand. Due to the size of these geckos it is often difficult to tell if females are gravid until a couple of weeks before egg-laying when the abdomen is obviously distended and the whitish outline of the fully formed eggs can be seen through the abdominal wall. The eggs are very large and will take up to 100 days to hatch depending on incubation temperature, which should be in the range of 27 to 29°C.

(Above) – **Male (left) and female Centralian knob-tailed geckos** *(Nephrurus amyae)* showing the size difference between the sexes.
(Below) – **Close-up of Centralian knob-tailed gecko** *(Nephrurus amyae).*

Keeping Geckos

GIANT CAVE GECKO
(Pseudothecadactylus cavaticus & lindneri)

This species and the closely related Kimberley giant cave gecko *(P.cavaticus)* are large natives of tropical northern Australia and thus require warmer temperatures in captivity than most other Australian species. Having said this they are still remarkably tolerant of low temperatures for short periods, even down to 10 to 12°C. Ideally, a hot spot of mid-30°C's should be provided by a coloured bulb or heat pad throughout most of the year. Over winter the temperature of this area can be reduced to 25 to 26°C while the remainder of the cage is allowed to drop to mid-teens. An enclosure set up similar to that described for the leaf-tailed geckos above can be provided, as these are agile climbers that require a tall cage design, e.g. 500x300x700mm high. Compatibility between male and female giant cave geckos appears to be the key to successful breeding. The female is often the more dominant of the pair and can become quite aggressive towards the male and cause significant injuries. They have powerful jaws and can bite extremely hard if handled incorrectly. Do not try to maintain these species as trios as the additional gecko, whether it be male or female, seems to cause friction between all cage mates and breeding is usually hindered and aggression is rife.

Once a compatible pair is established, they should breed every year but often only one clutch will be produced each season. Why this is the case is unsure. In *P.cavaticus* the eggs are laid quite early in the season, usually around October, so there is plenty of time for additional clutches, yet the single clutch appears to be the rule. The eggs are large and therefore do require a major investment by the female, but cave geckos are a large species with prodigious appetites so one would think they would have the ability to multi-clutch in good seasons of plentiful food. The eggs are laid in large plastic containers sufficient in size for the female to enter through a hole cut in the top and turn around easily once inside. The container can be filled with a moist substrate such as sphagnum moss or a sand/palm peat mix to a depth of 70 to 80mm. Incubation will take around 65 to 75 days at 28 to 29°C.

Australian Reptile Keeper

Kimberley giant cave gecko *(P.seudothecadactylus cavaticus).*

MARBLED VELVET GECKO
(Oedura marmorata)

This a an extremely variable and often very beautiful species that occurs from northern New South Wales up through Queensland to much of the Northern Territory. Strangely, there is an isolated population that is also quite widespread in the far west of Western Australia. The general pattern usually consists of bands of various colours around the body and tail but some populations are mostly spotted and there are various combinations of these markings that also occur. As with many velvet geckos the hatchlings are completely different to the adults with most colour forms producing babies which have a simple pattern of a dark background with broad pale yellow or cream stripes, often with small orange markings around the eyes. They are one of the largest of the velvet geckos reaching well over 100mm in snout-vent length in the largest specimens, although again there is a great deal of variation in size between localities.

Their large size requires plenty of space in captivity. To house 1-3 animals an enclosure of around 500x300x600mm high would be ideal. Never house males together if females are present as this species can be quite aggressive. They are capable climbers so provide plenty of branches, pieces of bark, etc. to enable the geckos to utilise all areas of the cage. A basking area can be provided as a small heat pad under one side of the base or a coloured light bulb in the top corner with a top temperature in the low 30's°C. During the day they will use hiding places either on or off the ground so rest layers of pieces of tree bark against the side of the cage and install other refuges on the floor. Marbled velvets are ravenous feeders and this behaviour can lead to obesity in captive animals. Offer food only two or three times per week with about 4-6 appropriately sized insects for each lizard. Gravid females and growing juveniles can be fed more often and their food should be supplemented with calcium and multivitamins on a more regular basis than non-breeding adults.

Females can produce numerous clutches each season, the eggs taking slightly longer to hatch than other velvet geckos at 75-80 days if incubated at 28-30°C. Hatchlings can be housed as small groups for their first 6-8 months but monitor them carefully to ensure that they are all receiving their fair share of food and none are lagging behind in their growth rates. The young geckos can be confidently sexed at around 10-12 months of age by which time it is essential that the males have been separated from each other if females are also present.

Australian Reptile Keeper

The marbled velvet gecko *(Oedura marmorata)* is one of the most variable species in Australia. Over its wide distribution there are numerous colour forms as well as differing tail shapes.

SOUTHERN SPOTTED VELVET GECKO
(Oedura tryoni)

The velvet geckos are some of the hardiest of all the Australian species and Tryon's or southern spotted velvet gecko is one of the easiest captives to care for. They are also an attractive species with an olive green background covered in dark edged pale spots, though the intensity of the pattern does vary somewhat with locality. Maximum total length is around 160mm. A pair or one male and two females can be adequately housed in an enclosure measuring 500x300x400mm high with a few hiding places, either natural or artificial and one or two climbing branches. A lid is essential as all velvet geckos are expert climbers of vertical surfaces even glass. Their natural distribution is the warm temperate to subtropical area of northern New South Wales and south-eastern Queensland, so maximum temperatures of around 32°C are ideal in summer dropping to mid- to low teens in winter. This cooling period may be quite important in this species to ensure successful breeding. Provide a small plastic container of sphagnum moss or similar that is kept permanently moist during the breeding season (October-February) and the females will lay their clutches of two eggs every 28 to 35 days. Incubation at 27 to 29°C will have the eggs hatching in approximately 60 to 65 days. The hatchlings will begin feeding at five to eight days and several can be safely housed together for the first six to eight months after which time it is important that the males be separated out from each other as soon as their hemipenal swellings become evident.

The velvet gecko group contain several other easy care and attractive species with similar captive needs to the southern spotted velvet gecko. The northern spotted velvet gecko *(O.coggeri)* and the northern velvet gecko *(O.castlenaui)* require similar housing but with slightly warmer temperatures to reflect their tropical distribution. Another fascinating species is the fringe-toed velvet gecko *(O.filicipoda)* which needs the warmer temperature but also a larger enclosure because of its size. A cage 500x300x700mm high would suit a pair or trio and provide a basking spot around 33-35°C. All of these species are beautifully marked and coloured and thrive in captivity living in excess of 15 years under the right conditions.

Southern spotted velvet gecko *(Oedura tryoni)*.

Keeping Geckos 72

SYDNEY BROAD-TAILED GECKO
(Phyllurus platurus)

A familiar species to Sydneysiders, this species has adapted to living on the walls of houses and other buildings, hunting at night for spiders, moths and other insects. In captivity they do well in an enclosure measuring 400x300x600mm with plenty of narrow vertical hiding places. The best furnishings to use for this set up are long strips of rigid bark placed diagonally from the front bottom of the enclosure to the back top. If four or five pieces are placed in this fashion on top of each other, a range of crevices of varying width is provided for the lizards. Although their natural habitat is sandstone, it is much safer to use bark to ensure that the geckos are not accidentally crushed between heavy pieces of rock. One or two additional climbing branches can also be supplied. A very gentle heat source can be provided at one end of the enclosure such as a low wattage, coloured light bulb at the top or small heat pad beneath the enclosure floor. The hottest part of the cage should not exceed 32ºC, as these geckos are very sensitive to heat. During the hottest days of summer the enclosure must be kept cool and if necessary sprayed with water several times during the day to keep the temperatures down. Exposure to temperatures over 35ºC for an extended period may lead to the death of these lizards. In contrast, they are very tolerant of low temperatures and, inhabiting a warm temperate climate, a winter cooling period with temperatures down to at least low teens is probably critical in ensuring reproductive success. Regular spraying of the enclosure with water should be part of the normal husbandry routine for this species; two or three times a week in summer down to once a week in winter. Try not to spray the lizards directly, as they appear to resent this by squeaking loudly and waving their tail back and forth.

Females will produce three to four clutches of eggs each season, which runs from September to February. Install a plastic sandwich box on the floor of the enclosure and fill this with a sand/palm peat mixture. Keep this constantly moist during the laying season and it will not only provide an egg-laying site for the females but will also contribute to maintaining the high humidity enclosure environment these geckos require. Like all leaf-tailed geckos, the appropriate incubation temperature is somewhat lower than that for most other Australian species. A basic temperature of 24 to 26ºC is ideal for this species, although slightly higher temperatures can be tolerated for short periods. At this temperature eggs should hatch in around 75 to 90 days. Hatchlings are quite tolerant of similar sized cage mates so several can be housed together for the first few months but keep a close eye to ensure they are all obtaining sufficient food.

(Above)- Sydney broad-tailed gecko *(Phyllurus platurus)* showing threat behaviour with the tail raised and the mouth open.

(Below)- Close-up of Sydney broad-tailed gecko *(Phyllurus platurus)* clearly showing raised scales and tubercules on the skin.

Keeping Geckos

ROUGH-THROATED LEAF-TAILED GECKO
(Saltuarius salebrosus)

Along with the northern leaf-tailed gecko *(S.cornutus)*, this species is the largest of the Australian geckos with snout-vent lengths of up to 145mm and a total length of around 250mm. The original tail is a spectacular flat, fleshy appendage beautifully ornamented with cryptic colours and textural tubercules and spines. In the event that the tail is lost it does regrow but is never as exquisitely marked and lacks any ornamentation. The rough-throated leaf-tail occurs in the sandstone cliffs and forests of a few scattered areas of central Queensland emerging at night to forage for large beetles, cockroaches and spiders. This distribution makes this species the most hardy of all the leaf-tails in captivity. It is capable of withstanding higher temperatures than all the other species including those occasional hot summer days of southern Australia where the mercury reaches the high 30's provided the geckos are not exposed to direct sunlight. Temperatures in excess of this may prove fatal over extended periods. A regular water spray of the enclosure during these times will assist in reducing the ambient levels by a degree or two. At other times a liberal spray once or twice a week is sufficient.

The enclosure should be set up as described above for the broad-tailed gecko, using the upright strips of rigid tree bark, but it should be substantially larger. Ideally two or three rough-throats should be housed in an enclosure around 600 x 400 x 800mm high. A heat source should be provided but needs to be no warmer than 32°C and isn't required over the winter nights when a cooling period should be offered. Standard commercially available insects make up the diet and can be supplied three times a week during the breeding season reducing to once every two to thre weeks in winter. These are long-lived and undemanding captives. Breeding is similar to that described for the broad-tailed gecko with a larger laying box and a slightly warmer incubation temperature of 26 to 28°C, although the eggs will still take up to 100 days to hatch. Juveniles can be easily raised in small containers measuring 150x150x250mm high set up in a similar way to the adult enclosure. Mist this container every second day and begin offering food about 10 days after hatching. Dust the food items regularly with calcium and multivitamin powder but only provide two to three prey items per lizard at each feed. These large geckos are slow growers and may take three to four years to reach maturity.

(Above)- Adult rough-throated leaf-tailed gecko *(Saltuarius salebrosus)*.
(Below)- Juvenile displaying tail-waving threat behaviour.

BYNOE'S GECKO
(Heteronotia binoei)

Although small (snout-vent length of only 50 to 60mm) and often more drably coloured than some of its relatives, the Bynoe's gecko is one of the hardiest of all Australian geckos. It occurs over a huge range both geographically and ecologically inhabiting many habitats over most of mainland Australia except the extreme south. Recent research suggests that it is actually made up of several closely related but distinct species, some of which are able to reproduce without the presence of males, a process known as parthenogenesis. This diversity is also reflected in the variability of the colour and pattern of these geckos over its range. Most of the available captive stock require both sexes for successful reproduction but some parthenogenetic specimens are available. They produce tiny brittle calcareous-shelled eggs at a prolific rate if conditions are good, with many clutches produced by a single female over a breeding season. Eggs can be incubated with no moisture content in the incubation medium and take around 60 to 70 days at 28°C. After hatching the babies grow rapidly on pinhead crickets and, if the food supply is maintained, they will breed in their first year. If cared for properly Bynoe's geckos have been recorded as living in excess of twelve years in captivity.

Being small in stature, their captive housing is correspondingly sized. An enclosure measuring 300x200x200mm high will adequately house two or three geckos. Provide three or four hiding areas plus a moist corner or small container of moist substrate for egg-laying and rehydration. A heat pad should be placed under one corner of the enclosure supplying a hot spot of no more than 35°C. These little geckos will rarely drink from a dish but it is a good idea to provide a small water receptacle if possible. Supplement this by spraying the enclosure lightly but regularly with water; once a day in summer reducing to once or twice a week in winter.

Bynoe's gecko *(Heteronotia binoei)*.

RING-TAILED GECKO
(Cyrtodactylus louisiadensis)

As a native of northern Queensland, the ring-tailed gecko requires a slightly warmer general temperature than many species that are commonly maintained. Having said this, they are not a delicate gecko and will tolerate quite cold temperatures during winter even down to single figures. Ideally, a heat pad should be provided all year round with a winter maximum of 27 to 28°C, which in summer can be increased to 36 to 38°C. They are attractively marked with pale narrow bands separating wider areas of dark brown. On the tail the bands become much narrower and paler, almost white in some specimens. There is a great deal of variation even over their relatively small range both in colour and size. Recent speculation has suggested that this variation may actually indicate that there may be more than one species in Australia. Some populations have recorded specimens over 300mm in total length, although around 240mm is the usual length. These are semi-arboreal and will use climbing branches and vertical pieces of bark, but spend much of their time on the ground. An enclosure of 750x300x500mm high would be ideal for one male and one or two females.

Ring-tailed geckos have a high demand for calcium, particularly for females during the breeding season. Their eggs are thick-shelled and heavily calcified and are produced at regular four to five week intervals over the warmer months. Both sexes have areas at the back of the mouth called endolymphatic sacs where calcium appears to be stored. These can sometimes become quite distended and visible externally as swollen areas on the lower neck. It is a good idea to provide a small dish of calcium with no additional multivitamins in the enclosure as well as regularly dusting all food items before they are offered to the geckos. The eggs are almost round in shape and approximately 12mm in diameter. They can be incubated in a dry medium at 28 to 29°C and will take up 140 days to hatch. The juveniles are more brightly coloured than the parents and will shed their skin within 24 hours of emerging from the egg. A small raising container should be set up with plenty of hiding places, as the hatchlings are very nervous and flighty for the first few months.

Ring-tailed gecko *(Cyrtodactylus louisiadensis).*

Keeping Geckos

COMMON SCALY-FOOT
(Pygopus lepidopodus)

This is a common species of legless lizard from southern Australia, although it does extend to the tropics on the eastern seaboard. It is highly variable both in colour and size with the largest specimens reaching a total length of up to a metre, almost three-quarters of which is tail. They naturally inhabit open woodland and heathland and may be found in sand dune country close to the coast in some areas. Mature males can be differentiated from females by their larger and more protruding limb flaps.

Their large length means a correspondingly large enclosure is necessary for successful housing. A cage of around 900x400x400mm is suitable for one or two adult scaly-foot. Ideally, they should be housed separately as they can be very aggressive towards each other and problems can arise during feeding. The females are often the more aggressive of the two sexes and the slightly smaller males can receive quite substantial injuries from an aggressive mate. When introducing specimens together for the first time they should be observed regularly for the first few days and checked often for the presence of bite marks and injuries. Scaly-foot love to rest curled in grass tussocks so these should be provided at all times. The grass doesn't have to be green and alive, dried clumps are just as suitable. Other refuges should also be provided, such as large pieces of flat bark or upturned pot saucers plus a dish of drinking water. A similar heat gradient to that described for the Burton's legless lizard is adequate for the scaly-foot. Scaly-foot are not fussy about their food and will accept most types of insects. It is advisable to offer food from forceps, especially with fast insects such as cockroaches as these are not always the quickest and most agile of hunters. These lizards also enjoy sweet soft fruit such as pear, grape, banana, etc. and a small dish of chopped pieces should be offered once a week. The fruit can be dusted liberally with calcium and multivitamin powder at each feed. An ultraviolet light source should also be installed in or above the enclosure as per the manufacturers recommendations. Captive breeding is very rare in this species. In many cases females become so aggressive towards the male that it is not possible for mating to be achieved. In other situations males and females may cohabit for many years with no reproductive activity taking place. Two eggs are laid in each clutch and females will produce at least three clutches in a season. Eggs take around 60-65 days to hatch at 27-29°C.

Australian Reptile Keeper

Common scaly-foot *(Pygopus lepidopodus).*

BURTON'S LEGLESS LIZARD
(Lialis burtonis)

This is the most widely maintained of all the pygopods in captivity. Even so, they become available very infrequently. The main problem with their maintenance is that weaning the lizards onto any other diet except other smaller lizards has rarely been achieved. Skinks are a particular favourite and while these food items are abundant in most gardens in Australia there are legislative concerns due to the protection of most reptile species in all states. Problems also arise with parasites infestation from the wild lizard prey, as well as the sheer inconvenience of having to capture food items on a regular basis. Skinks are usually offered only once or twice per week, more often for gravid females or growing juveniles, the latter obviously requiring appropriately sized prey items.

The feeding problem aside, Burton's legless lizards are relatively easy to maintain and make interesting and attractive captives. Some specimens are beautifully marked with longitudinal stripes of chocolate brown or white especially along the flanks, while others are pale grey-brown and almost patternless. An enclosure of 700x300x300mm is suitable for two or three lizards and a secure, tight-fitting lid is essential with all pygopods. A thin layer of substrate is needed along with several hiding places, placed at various points along the enclosure's temperature gradient. Like most legless lizards they often prefer to hide amongst grass tussocks so dried clumps should be provided where possible. Provide a hot area of 34 to 36°C using a spot-light or under floor heat pad with the coolest end of the cage set at around 24 to 26°C. As consumers of vertebrate prey, it is possible that these reptiles obtain all their vitamin D3 needs from their food. However, it may be beneficial to provide an ultraviolet light source as these are primarily diurnally active lizards and exposure to the rays of the appropriate wavelength may offer additional benefits. Full spectrum lights emitting both UVA and UVB would be ideal and are now readily available from pet stores. Breeding has been achieved occasionally with this species but very little information is available on this matter. As with their gecko relatives, two eggs are laid in each clutch and it is important that a moist, humid area is available in the enclosure during spring and summer as a potential egg-laying site for the female.

Burton's legless lizard *(Lialis burtonis).*

REFERENCES

Banks,C. Hawkes,T., Birkett,J. & Vincent,M. (2000) Captive Management & Breeding of the Striped Legless Lizard, *Delma impar*, at Melbourne Zoo, Herpetofauna 29(2):18-30.

Brown,D. (2005) Care & Husbandry of *Diplodactylus* Geckos. Part 1: Terrestrial Species, Reptiles Australia 2(4):32-36.

Brown,D. (2006) The Care & Husbandry of *Diplodactylus* Geckos Part 2: Arboreal Species, Reptiles Australia 2(5):12-18.

Deutscher,J. (2005) Captive Maintenance & Reproduction of Bynoe's Gecko *Heteronotia binoei*, Monitor 14(1):7-8.

Henkel,F-W. & Schmidt,W. (1995) Geckos, Kreiger, Fla., 237pp.

Husband,G. (1998) Captive Maintenance & Breeding of the Spiny-Tailed Gecko *Diplodactylus ciliaris* at Territory Wildlife Park, Dactylus 3(3):117-120.

Laube,A. (2000) The Soft-spined Gecko *Strophurus williamsi* Kluge, 1963, Gekko 1(2):30-35.

Laube,A. (2001) Captive Maintenance & Breeding of Some Ground Dwelling Australian Geckos. Part 1. *Nephrurus laevissimus* Mertens, 1958 & *Nephrurus levis* De Vis, 1886, Gekko 2(1):30-36.

Laube,A. (2002a) Captive Maintenance & Breeding of Some Ground Dwelling Australian Geckoo. Part 2.*Diplodactylus byrnei* Lucas & Frost 1896, *D.tessellatus* (Gunther, 1875) & *D.steindachneri* Boulenger, 1885, Gekko 2(2):12-18.

Laube,A. (2002b) Captive Maintenance & Breeding of Some Ground Dwelling Australian Geckos Part 3: *Nephrurus amyae* Couper, 1994, Gekko 3(1):43-48.

Laube,A. & Porter,R. (2004) Captive Maintenance & Breeding of Some Ground Dwelling Australian Geckos. Part 4:*Underwoodisaurus milii* (Bory de Saint-Vincent,1823) & *U.sphyrurus* (Ogilby, 1892), Gekko 4(1):23-32.

Porter,R. (1998) Captive Breeding & Maintenance of Tryon's Velvet Gecko *(Oedura tryoni)*, Dactylus 3(3):131-137.

Porter,R. (1999) Captive Maintenance & Breeding of Australian Leaf-Tailed Geckos *(Saltuarius & Phyllurus)*, Herpetofauna 29(2):13-17.

Porter,R. (2000) The Northern Leaf-Tailed Gecko Captive Maintenance & Breeding, Reptilia 13:50-55.

Porter,R. (2001) Captive Maintenance & Breeding of the Golden-Tailed Gecko, *Diplodactylus taenicauda* de Vis, 1886, Gekko 2(1):2-5.

Porter,R. (2003) The Fringe-Toed Velvet Gecko *Oedura filicipoda* King, 1984, Gekko 3(2):9-16.

Porter,R. (2004) Care of Australian Geckos & Legless Lizards, CARA Conference Proceedings:37-43.

Porter,R. (2005) Captive Breeding & Maintenance of Rough Knob-Tailed Geckos, Reptiles Australia 2(3):6-10.

Sonnemann,N. (1995) Captive Maintenance & Breeding of the Giant Cave Gecko *P.seudothecadactylus lindneri lindneri*, Monitor 7(2):66-76.

Sonnemann,N. (1998) Captive Breeding of the Giant Cave Gecko, *P.seudothecadactylus lindneri lindneri* (Cogger 1975), Dactylus 3(3):103-114.

Tremper,P.A. (1999) Captive Husbandry & Propagation of the Australian Eyelash Gecko *Diplodactylus ciliaris ciliaris* Boulenger, 1885, Gekko 1(1):10-15.

Wagner,E. (1998) Australian Velvet Geckos, Reptiles Feb.:36.

Wagner,E. (1999) Australian Geckos, Reptiles 7(10):38-39.

Wilson, S. & Swan, G. (2008) A Complete Guide to Reptiles of Australia. Second Edition. New Hollan, Sydney 512pp.